Eco-Trauma Cinema

Film has taken a powerful position alongside the global environmental movement, from didactic documentaries to the fantasy pleasures of commercial franchises. This book investigates in particular film's complex role in representing ecological traumas. Eco-trauma cinema represents the harm we, as humans, inflict upon our natural surroundings, or the injuries we sustain from nature in its unforgiving iterations. The term encompasses both circumstances because these seemingly distinct instances of ecological harm are often related and even symbiotic: The traumas we perpetuate in an ecosystem through pollution and unsustainable resource management inevitably return to harm us.

Contributors to this volume engage with eco-trauma cinema in its three general forms: accounts of people who are traumatized by the natural world, narratives that represent people or social processes which traumatize the environment or its species and stories that depict the aftermath of ecological catastrophe. The films they examine represent a central challenge of our age: to overcome our disavowal of environmental crises, to reflect on the unsavoury forces reshaping the planet's ecosystems and to restructure the mechanisms responsible for the state of the earth.

Anil Narine is a junior faculty member in the Department of Visual Studies and the Institute of Communication, Culture, Information and Technology at the University of Toronto. He completed graduate studies in the School of Communication at Simon Fraser University, and in 2011–2012 he was a Postdoctoral Fellow in Film at Columbia University. His research examines network theory and trauma theory in the context of globalization and thickening global connections. His publications appear in *Communication, Culture & Critique*, *Critical Studies in Media Communication*, the *Journal of American Studies*, *Americana*, *Memory Studies* and *Theory, Culture & Society*.

Routledge Advances in Film Studies

Eco-Trauma Cinema

Edited by Anil Narine

Routledge
Taylor & Francis Group

LONDON AND NEW YORK

First published 2015
by Routledge

2 Park Square, Milton Park, Abingdon, Oxfordshire OX14 4RN
711 Third Avenue, New York, NY 10017

*Routledge is an imprint of the Taylor & Francis Group,
an informa business*

First issued in paperback 2018

Library of Congress Cataloging-in-Publication Data

Eco-trauma cinema / edited by Anil Narine.
    pages cm — (Routledge advances in film studies ; 33)
  Includes bibliographical references and index.
  1. Nature in motion pictures.   2. Ecology in motion
pictures.   3. Environmental protection and motion pictures.   I. Narine,
Anil, editor.
  PN1995.9.N38E26 2014
  791.43'66—dc23
  2014016191

ISBN: 978-1-138-79139-8 (hbk)
ISBN: 978-1-138-54841-1 (pbk)

Typeset in Sabon
by Apex Covantage, LLC

To my parents, for showing me so many countries and so many movies.

Contents

Figures

All images appear according to the Four Factors of Fair Use, herein for the purposes of commentary and criticism, and education.

Acknowledgements

If this book were a film, there would be no close-ups. There would only be group' shots of an ensemble cast. Academic research projects are collaborative efforts, even when the publications have one name above the title. Anthologies like this one are an intensified form of collective labour. I would like to thank the contributors to this volume for their inspiring chapters, which went far beyond my expectations when I proposed the book and have taught me more than I thought I could know about this topic. Indeed, it is a remarkable thing that one person can express a term, or concept, or way of looking at things, and find colleagues across the globe who were thinking the same thing.

I offer gratitude to our editor at Routledge, Felisa Salvago-Keyes, who shepherded the book along from its earliest stages. Chuck Wolfe and Edward Branigan expressed enthusiasm about the book idea early on and motivated me to pursue it.

I am indebted to Gary McCarron, who supervised my thesis on cinema and social networks at Simon Fraser University, and to Shane Gunster and Andrew Feenberg, my committee members. Each had different insights into environmental crises that stayed with me, even as I developed this project without the luxury of their oversight. Douglas Kellner, my external examiner, became a great supporter. Jane Gaines, who was my Postdoctoral supervisor at Columbia University, offered support for several of my early, incubating projects. A large proportion of *Eco-Trauma Cinema* was edited in the Butler Library and the New York Public Library.

My department chairs, Anthony Wensley, Louis Kaplan, and Alison Syme have supported my research, teaching and travel endeavors with vigor, funding and kind, inspirational words. My colleagues in the Department of Visual Studies, Jill Caskey, John Ricco, Kajri Jain, Evonne Levy, Cindy Mallory, Meghan Sutherland, Brian Price, and Harriet Sonne de Torrens have all offered wise counsel. In the Institute of Communication, Culture, Information and Technology, Lisa Peden, Rose Antonio, Brett Caraway, Divya Maharajh, Rhonda McEwan, Tracy Bowen, Diane Pracin and Guy Allen have created a workplace and intellectual environment anyone would envy.

Acknowledgements

Introduction
Eco-Trauma Cinema

Anil Narine

Whether ecological catastrophes confront us directly, as experiences, or indirectly, as images circulating in the media, these events tend to confound us, stifle us and even paralyze us politically and psychologically. Nature, whether it threatens us, we threaten it or we see ourselves as part of it, remains sublime in this way: something too vast in its beauty and power to comprehend. This incomprehensibility may cultivate awe, but it may also hinder pragmatic human responses to ecological crises. In industrialized economies, media reports and documentary footage of diminishing glaciers, oil spills and deforestation now circulate widely, but they elicit tentative responses from viewers. Public discussions of environmental issues are intended to inform us as citizens, and the core of citizenship is the ability to take political action in the public sphere. But often these media representations of environmental crises can induce a sense of passive resignation, a sentiment that seems to be spreading. Recent studies suggest that North Americans are slightly less concerned about environmental issues than we were leading up to the founding of Earth Day and the Clean Air Act in 1970 (Tuttle 2012). Such polls are notoriously difficult to interpret in terms of causality, but two key changes that have taken place in the media during the past generation are surely factors. First, we have witnessed the rise of twenty-four-hour news stations, speciality media such as the Discovery Channel, the National Geographic Channel, Animal Planet, Disney Nature, BBC Earth, the globally celebrated *Planet Earth* (2006) and *Blue Planet* (2001) series and the proliferation of natural disaster blockbusters and ecologically themed fantasy franchises from Hollywood, Asia and Europe (Cubitt 2005; Murray and Heumann 2009). Second, given the array of choices they represent, these media compete fervently with one another for viewers and revenue, conjuring the most sensationalist and even shocking footage of human-induced ecological crises and people in peril that their producers can assemble. In sheer amount and tone, the very narratives about nature that can enliven our *best* political and social activities, aimed at protecting our imperilled planet, may also be so blinding in their intensity that they overwhelm us and prevent any activity whatsoever. This experience is a facet of *eco-trauma*, a concept this book investigates.

WHAT IS ECO-TRAUMA?

Eco-trauma results from a paradox that characterizes our age of anxiety. We *know* our ecosystem is imperilled, but we respond in contradictory ways. On the one hand, we want to take action to protect the natural world. Even though the environment seems to have become less urgent as a political issue, when compared with contemporary economic woes, people continue to take action to reduce their ecological footprints. On the other hand, it is also undeniable that we disavow our knowledge of climate change and dwindling natural resources in order to function more happily in a global economic context replete with unsustainable practices. According to this second line of thinking, we treat ecological harm as a trauma: something acknowledgeable that we work to repress in order to avoid its painful effects. As psychologist Tina Amorok theorizes in her study "The Eco-Trauma and Eco-Recovery of Being," "We defend ourselves from this fearsome side of inter-connectedness through separation ideologies and practices (war, religious fanaticism, racism, and sexism)" in addition to "psychological defense mechanisms (denial, dissociation, psychic numbing), and an array of debilitating behaviors and responses that bear the signature of trauma" (29). This psychoanalytic account of disavowal makes a convincing case for citizens' widespread inaction, because it highlights the unconscious *and* conscious nature of our paralysis; we are not simply passive, but actively retreat from uncomfortable realities. Disavowal was perhaps most famously described by Jacques Lacan as "Je sais bienmaisquandmême [I know very well, but nonetheless . . .]" (see Mannoni 1969). The paradox of eco-trauma encompasses non-psychological factors as well, including our physical proximities to imperilled ecosystems and our widely varying levels of economic influence—issues explored in the following pages. In seeking meaningful ecological action, and endeavouring to escape this responsibility, we experience twin desires pulling us in opposite directions, and disavowal is thus only one element of the wider problematic.

In our quest for a refuge from ecological catastrophes, and our quest to combat them, we can face a sensory overload—an experience also consistent with psychological trauma. But according to which definition? Trauma, in its modern theoretical manifestation, has a history reaching back to Victorian medical research on hysteria (Masson 1984). Freud redefined the concept several times throughout his life, as his research migrated from melancholic patients to soldiers. He outlined trauma vaguely in his essay "Beyond the Pleasure Principle" (1920) as "any excitations from outside which are powerful enough to break through the protective shield" of the ego (29). And the formation of Trauma Studies in the humanities and social sciences during the past three decades has only added further, nuanced definitions of this already difficult category of human experience. Even if they were to agree on a definition, many commentators still question whether trauma as a primarily *individual* experience can describe a society-wide experience (Radstone 2000; Sturken 1997; Walker 2005). Both issues are important to address here, especially since our collective project throughout this volume is to add another difficult concept—ecology—to discussions of trauma in the media.

What is trauma? According to Judith Herman's widely accepted psychiatric definition, "traumatic events generally involve threats to life or bodily integrity, or a close personal encounter with violence and death" (33). These events are beyond the realm of everyday experience because they overwhelm our faculties, "call into question basic human relationships" and "shatter the construction of self that is formed and sustained in relation to others" (51). Herman's book, *Trauma and Recovery* (1992), investigates a range of traumatic experiences, outlining the elements that unite and distinguish them, and yet trauma remains a difficult concept. The author, who has been extensively cited and validated for her leading research, has also been critiqued (Suleiman 2007). A key issue, and site of confusion, is trauma's *referent:* Trauma characterizes, first, the traumatic event; second, the victim's response; and third, his or her ensuing condition. Commentators must be precise, then, when referring to one of these facets of the traumatic experience. Once the event has taken place, the victim furthermore experiences trauma as an *absence*, presumably because the human sensorium protects itself by attempting to tune out the traumatic stimulus. Nightmares, flashbacks and phobias are a testament to the inadequacy of these protective psychic mechanisms; ultimately, the trauma has to be retroactively registered, in order for the patient to heal. This process is retrospective because, as Shoshana Felman and Dori Laub note in their study *Testimony: Crises of Witnessing in Literature, Psychoanalysis and History* (1991), "trauma precludes its registration; the observing and recording mechanisms of the human mind are temporarily knocked out" (72). Trauma is thus confounding in its definition and its lived experience.

Another site of inquiry is the question of who, exactly, is traumatized by cataclysmic events. According to Herman's description, only a victim who directly experienced the original event would qualify as traumatized. Indeed, the fourth edition of the *Diagnostic and Statistical Manual of Mental Disorders* (*DSM-IV*; American Psychiatric Association 1994) specifies that "natural disasters" are a common cause of psychological trauma, but only for those victimized first hand. This definition makes intuitive sense: Those injured and faced with deadly circumstances themselves can emerge traumatized. Others who bear witness to the traumatic event, according to this account, do not undergo its traumatic effects. Still, a range of scholars assert that *varieties* of trauma exist, challenging the above definition (Caruth 1995; Felman and Laub 1991; Kaplan 2008). Witnesses and even media viewers report feeling traumatized after seeing catastrophes unfold in their midst or on screen. How do we as media scholars address this version of mediated trauma? And how do these variations of trauma—which circulate among masses of viewers and *throughout society*—differ from descriptions of trauma in the individual?

A formulation of trauma in its socially mediated forms is necessary. If trauma is something that can afflict large groups of people not directly victimized by the event, its specific character needs to be articulated. E. Ann Kaplan does just this type of work, highlighting in her examination of images of catastrophe and their audiences how differing levels of trauma can circulate depending upon one's position as victim, bystander or witness. Kaplan posits these five proximities to trauma:

1. Direct experience of trauma (trauma victim).
2. Relative or close friend of trauma victim or clinical worker brought in to help the victim (close but one step removed from direct experience).
3. Direct observation by a bystander of another's trauma (also one step removed).
4. Clinician hearing a patient's trauma narrative—a complex position with both visual and semantic channels; it involves the face-to-face encounter with the survivor or the bystander within the intimacy of the counselling session (also one step removed).
5. Visually and verbally mediated trauma, that is, viewing trauma on film or other media, or reading a trauma narrative and constructing visual images from semantic data (two steps removed). (Kaplan 3)

When we see traumatic events represented in the mass media, including fictionalized accounts or re-creations, we are not true "witnesses" but rather viewers of visually mediated trauma. We are two or more steps removed from the "event" and thus cannot claim to be traumatized in the same way a victim or physically present witness may be. Still, as anyone who has wanted to "unsee" a photograph from the Holocaust or a news report of the recent genocide in Sudan can attest, viewing images of "distant suffering" can be traumatic. As far back as Freud's famous studies of "shell-shocked" World War I soldiers with his colleague Josef Breuer, definitions of trauma account for these varying intensities. According to Susan Rubin Suleiman, neurologists define a traumatic event as one that "produces an excess of external stimuli and a corresponding excess of excitation of the brain" which we are "not able to fully assimilate or 'process'" (276). For the victim or witness, this event would be the most intensely felt. But for the media viewer or even reader, a similar—though less intense—feeling of sensory overload may occur. Indeed, when faced with something unimaginable, our sensorium "responds through various mechanisms such as psychological numbing, or shutting down of normal emotional responses" (Suleiman 276). This propensity to shut down explains the difficulty many of us have "cognitively mapping" current or impending ecological crises (Jameson 1988). This challenge does not mean we fail to *interpret* these events entirely.

Indeed, human beings make meaning almost instinctively. We want to read simple signs as wonders, and when the signs are out of the ordinary and even traumatic in their intensity, we can interpret them as profoundly meaningful—and even as spiritual interventions in our lives. In such cases, we engage in what Žižek calls "the temptation of meaning," which is a psychological response to the chaos an ecological catastrophe reveals about our universe (2009, 158). As a way of coping, we impose meaning and even narrative formations onto chaotic events: New Orleans flooded and Las Vegas is drying up as punishment for the sins committed there, for example. In other words, trauma can produce a third type of response: First, we want to combat the trauma but relent because we feel overwhelmed by its magnitude; second, we want to disavow the trauma; and third, *we want to make meaning* from traumatic events, primarily as a coping strategy. It is crucial to note that this meaning may not coincide with the actual causal factors underpinning a given trauma. Eco-trauma encompasses all three of these responses. It is an important term to add to the discussion of ecological cultural politics because all three ways of responding to overwhelming stimuli *lead us astray* from effective preventative and restorative action.

ECO-MEDIA AND ECO-CRITICISM

Nowhere is the temptation of meaning more evident than in the stories assembled by the news media. When depictions of natural disasters circulate, they assume the ritualized forms of amateur footage of hurricanes, wildfires, flooding and displaced people, or satellite imagery of devastated plains in Africa and the Americas. In the West, these ecological events and processes (even when they may be incited by human practices) are designated as Acts of God. And in their aftermath, stories of survival similarly unfold in Biblical terms, as miracles, or God's will, bringing order back to the universe. And when stories of human-induced ecological disasters such as oil spills find their way into the mass media, they are similarly rendered meaningful, as *morality* tales. The accounts often play out according to the tested narrative of corporate greed versus community welfare. Ethically vacuous villains conspire in luxurious boardrooms while earnest citizens band together in the David versus Goliath narrative epitomized by the reporting on environmental activist Erin Brockovich's discovery of poisoned water in Hinckley, California, and her ensuing class action lawsuit. Through the victims' collective recuperation in the aftermath of an ecological trauma, social bonds are reaffirmed and the world regains its meaning.

In many cases, the news media earnestly wish to convey the urgency of the ecological crises they cover. These endeavours presumably aim to make viewers confront events that are remote from their immediate experience. Too often, however, sensationalism and the aforementioned narrative conventions take the reins, and citizens are presented with the familiar story of a polluter brought to justice, as if the infraction were an isolated incident or something activists have already addressed (Brockington 2008). Nuances, that is, are often lost in the age of the news media sound bite. To acknowledge this truism is not to condemn traditional journalism but rather to illuminate the wider political economy of the contemporary mass media, wherein companies compete with one another—as well as with citizen journalists and social media—for our momentary attention. News media are fortunately not the only public forum in which contemporary or impending ecological crises are represented.

Literature, the visual arts and film—even canonical texts not considered "environmentalist"—have been grappling with major questions about nature's fragility for generations. Eco-criticism (Buell 1996; Cubitt 2005; Garrard 2004; Glotfelty and Fromm 1996; Heise 1995) has grown as a field since the 1990s and illustrated that sophisticated inquiries into environmental concerns play out in these unexpected creative forums. In learned organizations such as the Association for the Study of Literature and the Environment (ASLE), scholars investigate a range of narrative forms. Works addressing literature from this perspective are numerous, with studies such as Lawrence Buell's *The Environmental Imagination: Thoreau, Nature Writing, and the Formation of American Culture* (1996) achieving canonical status, and more recent, theoretically informed inquiries such as Ursula K. Heise's *Sense of Place and Sense of Planet: The Environmental Imagination of the Global* (2008) rising to prominence in literary studies and beyond. While this bibliography is too voluminous and far afield to detail here, film eco-criticism has a shorter history.

From examinations of nature as a utopian space in the cinema, to rhetorical analyses of documentary films on climate change, various studies of eco-cinema helped shape this one. Adrian J. Ivakhiv's *Ecologies of the Moving Image* (2013) examines cinema's ability to elicit affective responses from representations (in film, digital formats and analogue animation) of environmental crises. Robin L. Murray and Joseph K. Heumann's *Ecology and Popular Film* (2009) addresses a variety of mainstream films, from pulpy special effects films to sombre dramas, suggesting that both high and low film forms disseminate influential representations of the environment. Murray and Heumann's *That's All Folks?* (2011) provides a comprehensive historical and critical evaluation of nature in US animation, from Disney to Warner Bros. The book contains a notable introduction that usefully historicizes the environmental movement in the US in the early twentieth century alongside the growth of cinema through the decades. Paula Willoquet-Maricondi's edited collection, *Framing the World* (2010), delves into complex eco-films, such Werner Herzog's *Grizzly Man* (2005) and major documentaries that have arguably changed the public discussion of ecological concerns in North America. *Ecocinema Theory and Practice* (2012), edited by Stephen Rust, Salma Monani and Sean Cubitt, assesses renderings of the natural world as a refuge, a panacea and a threat, from the dramatization of returning to nature in Sean Penn's *Into the Wild* (2007) to the shock of seclusion in 1970s horror films such as Wes Craven's *The Hills Have Eyes* (1977). Sean Cubitt's *EcoMedia* (2005) also assesses a range of recent films, some of them unexpected titles such as *X-Men* (2000) and *Whale Rider* (2002), which receive compelling readings. In his book *Hollywood Utopia* (2005), Pat Brereton evaluates films like *The Yearling* (1946), *Jaws* (1975), *The Emerald Forest* (1985), *Jurassic Park* (1993) and *Waterworld* (1995) in terms of the utopian and menacing visions they present of familiar landscapes, and ecosystems where humans are out of place. Scott MacDonald's *The Garden in the Machine* (2001) provides an expansive investigation of avant-garde films from Marjorie Keller, Claude Lanzmann, Stan Brakhage and others; MacDonald reads these films in unexpected ways, drawing out an ecological unconscious in works that presumably address quite distant issues such as the Holocaust and questions of film's materiality. Derek Bouse's *Wildlife Films* (2000) is a history of the nature documentary, which advances an argument about the films' and viewers' propensity to want to view animals as protagonists and nature as analogous to society with threats and codes or morality. In *Green Screens* (2000), David Ingram examines major issues in the visual representation of nature, such as renderings of landscapes on film, the frontier, nuclear power as a threat, myths of purity and Amazonian representations in the cinema. In *Reel Nature* (1999), Greg Mitman traces the history of the wildlife documentary in the US and argues that filmmakers are less concerned with the vérité of what they film than they are with imposing anthropocentric narrative patterns onto the imagery, which ensure an entertaining spectacle. Jhan Hochman in his book, *Green Cultural Studies* (1998), examines the age-old division between "culture" and "nature" and argues that our tendency to separate them enables the abuse of the natural world and the mistreatment of people, grouped into categories of race, class and gender. Whereas thinkers such as Donna Haraway and Andrew Ross advocate blurring the distinctions between animal and human, nature and technology, Hochman shows these approaches to reaffirm the promotion of culture over the value of nature as an end in itself. Alison Anderson's *Media, Culture and the Environment* (1997) similarly looks into our domination of nature, and ensuing natural and industrial disasters, but this inquiry focuses less on popular representations of ecology in cultural forms and more on news media framing, journalistic discourses and ideology.

Eco-criticism is a clearly a dynamic branch of contemporary critical theory. It makes undeniably clear that popular media, eco-themed literature (Kerber 2010) and commercial and documentary film have played central roles within the contemporary environmental movement since its inception in the 1960s. As the current renaissance in documentary film production and consumption shows, viewers and readers often turn to narrative forms in the hope that they can map social problems and demystify the realms of politics and science. Rachel Carson's nonfiction book *Silent Spring* (1962) is an example of an ecologically themed narrative that became a rallying point for this movement, lauded as *the* foundational text (Minster 38n). Carson made a powerful plea for a re-assessment of nature's sanctity, underlining how the everyday chemicals we use violate delicate natural processes. Carson's contribution had an astonishing impact on public perceptions of esoteric ecological dilemmas. According the Ursula K. Heise, the book is a unique combination of "pastoral" renderings of living things in peril, on the one hand, and blunt scientific analysis of chemistry, agri-business and pesticides such as DDT, on the other (2006, 512). *Silent Spring* likely took hold of concerned citizens' imaginations because it presented a *possible* future, a spring with no birds or insects, that was evocative and compelling in ways scientific reporting strains to be. Opening with "images of natural beauty" and highlighting "the 'harmony' of humanity and nature that 'once' existed," *Silent Spring*'s "pastoral peace rapidly gives way to catastrophic destruction" (Garrard 1). Carson's persuasive power therefore hinged on her engagement of readers' logical and emotional sides, helping us imagine the type of traumatic life world that many people at the time found unimaginable. Carson's rhetorical force drew upon her impressive scientific research, but her book's popular impact resulted from her *creative* rendering of a possible world.

Various cultural forms help us understand environmental crises. Other, more overtly fictional literary works, such as Richard Matheson's *I Am Legend* (1954), James Dickey's *Deliverance* (1970), Leslie Marmon Silko's *Almanac of the Dead* (1991), Margaret Atwood's *Oryx and Crake* (2003), Cormac McCarthy's *The Road* (2006) and even Suzanne Collins's publishing and Hollywood goldmine *The Hunger Games* (2008, 2009, 2010) examine struggles over human health, environmental protection, land and resources. On television, speculative nature programs such as the History Channel's *Life after People* (2008–2011) conceive of versions of the apocalypse in which human life ceases, forests replenish and new creatures emerge as the earth recovers from our presence. David Suzuki's *The Nature of Things* (1960–present); David Attenborough's *Life* series (1975–2005), all re-released in 2005; *Blue Planet* (2001); *Are We Changing Planet Earth?* (2006); and *Planet Earth* (2006) have found massive national audiences. Varying in subtlety, these programs suggest that human consumption habits are often unsustainable, whether it is Attenborough addressing the birds killed for colourful ink by local tribes in New Guinea or David Suzuki lamenting the unsustainable design of the Canadian "mega-city," Toronto. In the proliferating world of digital gaming, now outpacing Hollywood in global profits, some of the most popular series such as *Resident Evil* (1996–present), *Half-Life* (2004–present) and *Gears of War* (2006–2011) concern protagonists struggling to survive in earthly and otherworldly ecosystems ravaged by toxins and environmental degradation. The landscapes in the nearly photorealistic games are often uncannily recognizable, as are the military-industrial solutions to social problems that the games orchestrate—relishing in but also condemning the violence they choreograph.

On one level there is surely a superficial pleasure in beholding the traumatized landscapes that fictional representations conjure. Unlike factual reports about global warming, creative renderings of ecological crises seem to attract readers and viewers. Something intrigues us about an empty New York City overgrown with weeds, or downtown Seoul being menaced by a creature produced from the city's pollutants. On a deeper level, such scenarios engage us seriously, and unnervingly, in the work of mapping real ecological crises and their unpredictable effects on us as social and ethical beings, perhaps in the way George Orwell's *1984* (1948) gripped readers with its depiction of earth as a wasteland governed—not coincidentally—by totalitarian rulers using the media to promote ongoing war. Such creative cultural forms, in their best incarnations, prompt us to consider our quickly evolving subject positions, characterized by oscillating feelings of agency and helplessness in the face of contemporary ecological traumas. The narratives that represent future ecological challenges often aim to do more than merely titillate viewers and readers. They prompt us, perhaps inadvertently, to "look awry" at *contemporary* ecological crises, whereas more "direct" representations may cause us to disengage (Žižek 1991; 2009). In other words, these representations highlight the "socially symbolic" function of narrative forms in shaping public perceptions of actually existing conditions (Jameson 1981). In cases where news media and even scientific findings may face challenges connecting with an over-stimulated public, art forms such as film may hold to key to mapping contemporary eco-traumas.

ECO-TRAUMA ON FILM

This book investigates film's complex role in representing ecological traumas. From didactic documentaries to the fantasy pleasures of commercial franchises, film has taken a powerful position alongside the global environmental movement. Eco-trauma cinema represents the harm we, as humans, inflict upon our natural surroundings, or the injuries we sustain from nature in its unforgiving iterations. The term encompasses both circumstances because these seemingly distinct instances of ecological harm are often related and even symbiotic: The traumas we perpetuate in an ecosystem through pollution and unsustainable resource management inevitably return to harm us. Eco-trauma cinema takes three general forms: (1) accounts of people who are traumatized by the natural world, (2) narratives that represent people or social processes which traumatize the environment or its species, and (3) stories that depict the aftermath of ecological catastrophe, often focusing on human trauma and survival endeavours without necessarily dramatizing the initial "event." Of course, for millennia Western and Eastern, South American and African cultural forms have examined the fragile balance between people and ecosystems. Similarly, art through the centuries from the time of the Egyptian pharaohs has engaged with our fears of the apocalypse. The eco-trauma culture investigated here is decidedly contemporary by contrast, even though these texts extend much earlier traditions. In world cinema ranging from Japan's *Godzilla* (1954), to the US thriller *Deliverance* (1972; see Narine 2008), to Australia's *Long Weekend* (1978) and German-American director Roland Emmerich's climate change spectacle *The Day after Tomorrow* (2004) and apocalyptic *2012* (2009), ecological devastation and ecological threats to human life have been major themes.

Eco-trauma films are often deceptively simple in narrative terms. In one variety, urbanites leave the city for an experience in a challenging natural landscape—such as the remote Appalachian river in *Deliverance*, the Australian Outback in *The Long Weekend* or the perilous frozen outposts in *The Thing* (1982; 2011), *The First Winter* (1982) and *The Last Winter* (2006). These protagonists often discover that modernity has stripped them of any familiarity with nature, limiting their ability to live according to its demands. In a second variation, protagonists set out for a natural experience in search of solace and even healing. A surprising number of these films involve couples seeking seclusion in nature in order to mend marriages (*The Shining* 1980; *Long Weekend*) or to grieve the loss of a child (*Dead Calm* 1989; *BABEL* 2006; *Antichrist* 2009). In *Vinyan* (2008) and *The Impossible* (2012), upper middle class European families opt to vacation in Thailand because of its natural beauty and tourism industry, never expecting the traumas awaiting them when the ocean swells to encompass their beach resorts. The first film follows its protagonists, grieving for their lost child, away from the resorts and into the wilderness of Southeast Asia. In a third incarnation, an external threat infiltrates an idealized community, destabilizing the social order and testing the values of each resident. Echoing various Cold War–era alien invasion thrillers, *Jaws* (1975) is the paradigm case (revamped countless times, notably in *The Host* 2006 and *Teeth* 2007), given that the police chief has relocated to the Capra-esque small town of Amity to escape the traumas of big city crime and the existential dilemma of risking death day after day on the job. The polysemous shark can thus be equated with this returning, repressed threat of violence or Vietnam War flashbacks, as well as with communism, voracious capitalism or the lingering traumas of war, which afflict the shark hunter, Quint, whose many comrades were killed by sharks while awaiting rescue following the sinking of the USS *Indianapolis* on July 30, 1945.

As *Jaws*'s memorable reference to this historical event illustrates, the line between fictional narratives and the real can be confusingly blurred in eco-trauma films. Our preliminary suspicion in this book is that the ecological harm we have caused, and the ecological threats that may await us, strike viewers as actually existing phenomena, and thus more menacing than any outlandish villain Hollywood might conjure. According to Jameson in his famous essay "Reification and Utopia in Mass Culture" (1979), the fantasy and the terror of *Jaws* was in fact a remarkable snapshot of American capitalism during that cultural moment: The working, middle, and ruling classes come together for a utopian—and tragically temporary—moment to destroy the external threat, at a time when an economic recession saw US industry ravaged by the infiltration of commodities from former enemies Germany and Japan. Jameson thus illustrates that invented imagery can evoke the trauma of the real; in the film, economic and social traumas were symbolically *embodied* in a single form—a leviathan—that could be obliterated therapeutically. Various anxieties took a singular form.

Throughout the summer of 1975, as record crowds lined up to see Steven Spielberg's shark opus, cinematic and non-cinematic worlds overlapped, and viewers found themselves responding to invented scenarios and illusory threats in their actual lives. *Jaws*'s depiction of the fantastical seaside world of Amity being menaced by a great white produced hysteria about shark attacks and (along with some highly publicized attacks in the 1970s) contributed to the global vilification of sharks. Mass culls and unethical amateur shark fishing took place across the coastlines of the US, South Africa and Australia. Paranoia about sharks has never quite subsided since the release of *Jaws*, even though bees kill more people annually. Only recently have courageous documentaries such as Rob Stewart's *Sharkwater* (2006) challenged the mindless killing of sharks for sport, as well as showcasing how many are killed as the collateral damage of drift nets. In parts of Canada and elsewhere, sharkfin soup—responsible for the majority of shark fishing—has been banned. But sharks face continuing threats, not merely from China and its demand for fins, but from national culls in Australia in 2013 and 2014. Clearly, cinema on the one hand, and perceptions of reality on the other, intermingle in complex ways. What better way to illustrate this point than to cite a fictional scientific authority? For as disease control specialist Dr. Mears (Kate Winslet) remarks in the Steven Soderbergh's eco-trauma film *Contagion* (2011), about a killer SARS-like virus, "a plastic shark kept millions of people out of the water. But a warning label on the side of a cigarette carton can't keep millions of people from smoking."

More recent cinema has provided similar instances of "real-world" ramifications. As Adrian Ivakhiv compellingly argues, we should not be surprised by this famous example because "Cinema . . . produces and discloses worlds. It is cosmomorphic: it provides for the morphogenesis, the coming into form, of worlds" (2). The cinematic world—now aided and sometimes hampered by easily malleable D-cinema formats and digital visual effects—becomes its own world, another reality *we begin to inhabit* with the full range of our senses, much like the paraplegic hero, Jake Sully, does in the speculative eco-trauma film *Avatar* (2009; see Bron Taylor's anthology on the film, Avatar *and Nature Spirituality* 2013). Indeed, after returning to their lives, following one or two viewings of *Avatar*, thousands of viewers wanted to "return" to the world of Pandora, where the action takes place. *Avatar* explores this sentiment in the character of Jake himself, a young man who, having lost the use of his legs, is exhilarated by his new life (and body) in Pandora's natural paradise (Figure 0.1). For their part, many viewers felt they had been to Pandora, surrounded by its 3-D vegetation, and they became depressed by the fact that Pandora did not exist beyond the cinematic frame (Sandler 2010).

Figure 0.1   Jake Sully (Sam Worthington), who uses a wheelchair, cannot resist going for a run on Pandora moments after being transferred biologically and cybernetically into his Na'vi body.

Therefore, although the documentaries and narrative films examined here receive distinct *analytical* treatment consistent with the discreet issues they raise as nonfiction and fiction, we hope to trouble—as Rachel Carson and Fredric Jameson do in different ways—any clear division between fiction and nonfiction, the real and the fantastical. What often rumbles beneath the surface of fictional cinematic tales, from the gruesome aftermath of the Asian tsunami depicted in *Vinyan* to the comedic duo in the digital blockbuster *WALL-E* (2008), are questions about the real world. These are existential quandaries about basic human needs, communal values, and cycles of human ignorance and self-destruction that have little to do with the threat presented by nature's fury. In fact, it is the nature of people, too often capable of devastating the natural environment and other living things, that comes most sharply into focus when disastrous events unfold. It should be unsurprising, then, that psychologists and psychoanalytically trained scholars in the natural and human sciences have intervened in the changing and problematic relationship between our environment and ourselves. A special issue of the psychology journal *Spring* was entitled "Environmental Disasters and Collective Trauma." Editors Stephen Foster and Nancy Cater compile analyses of the Fukushima nuclear crisis, following the 2011 tsunami, alongside readings of films such as *Melancholia* (2011) and *Out of the Ruins* (1989). Thus, whether protagonists are threatened by an unfamiliar wilderness, having left home, or by an intruder whilst safe in their community, with its sheriff and mayor standing guard, as in *Jaws* and the very real Hurricane Katrina disaster, the trauma that begins to unfold often tests the social bonds that in normal circumstances enable a shaky harmony between people of different backgrounds and social classes.

In the films we address, it is clear that a traumatized earth begets traumatized people. Nature, as we see in each author's excavation of these thoughtful films, does not simply enact its revenge upon us. Rather, it sustains and endures trauma as a human victim would, only its scarred body can never be made to disappear. In this way, through its ever presence, and the ubiquity of its scars, nature does not merely visit our own violence back upon us, but rather forces us to confront our own *propensity* to inflict the traumas to which ecological degradation, like scar tissue, bears witness. Witnessing is perhaps cinema's highest purpose. And in this way, as film theorist Bill Nichols argues, every film is a documentary—if not of actual people or events, then of a time, its automobiles, hairstyles, developed and undeveloped spaces. In the book's opening chapter, Barbara Creed examines a documentary and a drama that both bear witness to our changing and increasingly troubled relationship with food.

The source of food is a political issue in our age of accelerated globalization. Amid this metastasizing economic system, driven by the cultivation of new markets, networks of supply, production and consumption link even the most distant regions. And thus *where* nations source their food has become as pressing a question as where we source our oil and other natural resources, one which is often inhibited by global networks of dissipated accountability. The problem of accountability figures unsettlingly in nearly every contemporary account of ecological trauma. Even when a major polluter leaves behind toxic wastelands, as Chevron did in the Ecuadorian Amazon and as BP did in the Gulf of Mexico, the roles played by other agents in the network come to the fore: The subcontractors are to blame; the local inspectors failed to maintain standards; the executives at corporate headquarters were unaware of the problems. As Barbara Creed illustrates in Chapter 1, "Evolution, Extinction and the Eco-Trauma Film," the documentary *Darwin's Nightmare* (2004) and the drama *A Zed & Two Naughts* (1985) in fact address the issue of human accountability on surprisingly similar moral terms. Hubert Sauper's *Darwin's Nightmare* takes viewers to the shores of Lake Victoria, Central Africa, where fishing villages somehow fail to feed their populations despite hauling in hundreds of Nile perch each week. The species, introduced to the lake by the British in the mid-twentieth century, has decimated the lake's other fish and grown grotesquely huge in the process—an ambivalent, Darwinian adaptation. Indeed, some of the fish outweigh the men who catch them. European demand for their fillets leads the villagers to sell the fish to the indifferent Indian owners of the local cannery, where underpaid workers feel fortunate for the pittance they receive. Locals boil fish bones and heads and other refuse in their attempts to feed themselves. Eastern European pilots descend on the villages, happy to fly tons of fish back to Europe and to support the sex trade in the Tanzanian villages they frequent. And yet, the film seems to ask, who can be held to account for this dystopian picture of globalization? Can the global "market" simply be the culprit, an invisible hand that makes the fish more valuable in Europe? Creed resists such platitudes and considers the human at the centre of these relations—a human who, ironically, should no longer consider itself the centre of all things. Indeed, Creed historicizes the contemporary, Western anthropocentric worldview in a way that highlights its radical divergence from Darwin's enlightened theories of a natural world competitively achieving balance. The Nile perch allegorizes the rapacious growth of the human population and appetites—a Darwinian nightmare— and the unprecedented dominion we now exert over the earth.

Similarly, from the opening moments in Peter Greenaway's drama, *A Zed & Two Naughts*, we see the human being's place in the world to be callous and out of touch with other living things. Children drag a suffering Dalmatian down the sidewalk, a man observes a tormented tiger in captivity pacing in its cage and motorists collide with and kill a swan. Human relationships are disrupted by these animal encounters, and thus people are at least made aware of their cruelty. In fact, the future of humanity is itself not assured in Greenaway's unusual narrative, and thus people are subject to harm as well. Two passengers in the car accident are killed, another loses a leg and an ensuing discussion—amongst these urbanites somehow living in the midst of animals—concerns how long a woman's body takes to decay and be resorbed by the earth. In both *Darwin's Nightmare* and *A Zed & Two Naughts* the human presence in the natural world is potentially traumatic to other living things. And yet, as both films also show, humans sustain traumas themselves, almost as evidence that their appetites are out of balance. Our ongoing misunderstanding of our place in the wider ecosystem should necessitate a deeper engagement with Darwin's seminal work in which the wonders of life in all its forms were celebrated for their ingenuity, and people were granted no exemptions—neither from the biological reality that we are animals ourselves, nor from the social reality that our ethical decisions will either ensure the demise or continuation of life on earth.

In Chapter 2, "Trauma, Truth, and the Environmental Documentary," Charles Musser provides a history of the present by examining more than a century's worth of nature films, treating them as cognitive maps of a changing environmental consciousness. Ecologically themed nonfiction films had complex origins. From their beginnings with the Lumiere brothers' *Lion, London Zoological Garden* (1895), films documenting the earth and its creatures did not emerge with intact conventions audiences could learn to anticipate. Instead, in the early decades of cinema, nature films took a wide range of cinematic forms, from wildlife documentaries with expository voice-over narration, to short films about national parks such as the Yosemite and Yellowstone, to semi-staged adventure documentaries such as *Hunting Big Game in Africa* (1909) and *Roosevelt in Africa* (1910). These nonfiction films also shared little common ground ideologically, at least during this period. On the one hand, they could celebrate a sacred space and promote the conservation of a rare species, while on the other hand film-makers could just as readily portray humans "taming" nature, and commend aggressive agricultural practices or imperialist hunting expeditions in Africa. The question of what nature *was*—a delicate system to be protected or a threat to settlers' survival—was unresolved in the films of the period, just as it was in the popular consciousness of the period.

In the 1920s and 1930s, some environmental films took the form of the social issue documentary, and shed equal light on the plight of precarious people, and the fragility of nature. In Pare Lorentz's masterpiece *The Plow That Broke the Plains* (1937), farmers' struggles and mismanaged agricultural land receive attention in a way that celebrates human persistence without being overtly anthropocentric. As Lorentz's farming documentary exemplifies, film shaped public perceptions of natural resource exploitation in ways distinct from newspapers, radio broadcasts and the other media of the period. Musser therefore works from the premise that wildlife documentaries were "an expansion of human vision, a means of entering into a world that was invisible to the human eye, an extension of the physical body of the subject, allowing for the creation of pleasure by bringing animals in their natural habitat closer to humans" (Horak 459). The environmental documentary's impact on public discussions of environmental issues cannot be measured, but Musser's history of the form helps us understand our present fascination with eco-documentaries. In these contemporary films, this book suggests, we are watching a horrible, traumatic accident unfold slowly and in countless realms, from the destroyed coral lining the floor of the ocean to the melting polar ice caps. Can any of us discuss global warming without *An Inconvenient Truth*'s time-elapsed images of shrinking glaciers coming to mind?

In Chapter 3, "Great Southern Wounds: The Trauma of Australian Cinema," Mark Steven examines the similarly expansive and potentially menacing contours of Australia's widely varying ecosystems. Populated for millennia by Aboriginal people, even the country's remotest territories are affected—as in any postcolonial country—by the history of the political violence wrought by Europeans in the years since their arrival in the 1780s. These spectres commingle with the related *ecological* traumas that have obliterated traditional practices across Australia, and upset the delicate relationship between Aboriginal people and early settlers, and the land that sustained them. Steven thus suggests that the relationship between ecology and human trauma is unusually close in Australian national discourses. As in the Arctic, the landscape down under can readily become a traumatizing adversary to the human visitors traversing its expanses, as hunters and poachers (*Wake in Fright* 1971; *Crocodile Dundee* 1986), benevolent young adventurers (*Picnic at Hanging Rock* 1975; *Wolf Creek* 2005) or well-meaning campers seeking refuge from city life. According to Steven, cinema is uniquely effective at rendering the devastating impact of the violence visitors inflict on various creatures and ecosystems, and the traumas people endure in the midst of Australia's inhospitable expanses. The docudrama, *A Cry in the Dark* (1988), is a paradigm case based on an actual event: A young couple vacationing near Ayres Rock have their lives upended when a Dingo drags their baby from their tent and disappears into the darkness. The initial trauma is merely the beginning of a crisis, fuelled by a national media frenzy, which would cut to the core of Australia's self-understanding. Was the couple negligent in taking their infant into such a menacing environment? Are such outdoor activities not central to Australian life? Was the mother really the murderer, a religious fundamentalist making a sacrificial offering and using Nature as a silent scapegoat for her infanticide? As Steven shows, Australian cinema often concerns ecological traumas that in turn illuminate the country's deep ideological divisions.

In Chapter 4, Alf Seegert redirects our attention from national discourses and ideology, offering instead a gripping study of individual psychology. In his study, entitled "Into the Wilde?" Seegert applies N. Katherine Hayles's term "technotext" to describe Werner Herzog's controversial documentary, *Grizzly Man* (2005). A technotext foregrounds the mode of its own making and the materiality of the medium from which it is made (Hayles 794). And in this documentary, those media are the videotape and digital film that grizzly bear enthusiast Timothy Treadwell uses to document his dozen summers among the bears near Kodiak, Alaska. The resulting film is meta-documentary. Herzog, assembling the footage left behind, shows us many of Treadwell's "outtakes" and rehearsals. But *more* than a film is being produced in the video and film we see. As Treadwell records bear activity and passionately interviews himself, he begins to fashion (and question) his very identity. As viewers, we begin to understand that the naturalist has come to Alaska to produce himself anew. Is he a crusader for the bears' survival? As the local authorities make clear, Treadwell is in fact an intruder in a vast nature preserve where poaching poses no great threat. Is Treadwell publicizing the bears' plight, or his own, as an extreme conservationist in search of fame? We learn that the amateur filmmaker struggled as an auditioning actor in Hollywood, before supposedly losing a role to Woody Harrelson and giving up on that dream. Indeed, he abandons "the human world" in its entirety, according to Herzog, who narrates the film like a psychologist providing a forensic character assessment. For Treadwell, alone for months on end, the camera becomes a mirror in which he restyles his image, a priest to whom he confesses and a conduit to a presumed audience that will admire him as a "kind warrior" fulfilling his calling in life: to "die for these animals" in the service of "educating the public."

Like Treadwell's presence among the grizzlies, people are often the "contaminant" in eco-trauma films, and if we have not affected wilderness regions with our relentless presence, our toxins likely have. In Chapter 5, "The Dangers of Bio-Security: *The Host* (2006) and the Geopolitics of Outbreak," Hsuan L. Hsu examines Bong-joo Park's creature film as a cautionary vision of industrial contamination in South Korea and its effects. At first glance creature films like *The Host* seem to vilify the animal, the non-human landscape, or the alien body. But in fact they often entice viewers to investigate the human sources of toxicity, mutations and contaminants; the creature is thus a human creation, and its vengeance is punishment for—and a perverse mirror image of—human ecological violence. This equation, however, is never quite as symmetrical as it appears because different human actors share different levels of responsibility for ecological degradation, in the films as in life. Indeed, sophisticated films like *The Host* raise disturbing questions about human encroachment upon sacred spaces, and industrial indifference to the fragility of ecosystems near crowded urban centres, by locating guilt in *various* places. For Hsu, *The Host* can be understood as a revision of the "outbreak narrative"—a term scholar Priscilla Wald uses to describe the epidemiological plot genre about the global circulation of a virus, toxin or plague. Often these plots depict medical authorities in poorer regions of the world as impotent and even culpable for the outbreaks they failed to contain. Western discourses about the Chinese-borne SARS pandemic illustrated this troubling process of vilification. *The Host* disperses guilt, presenting a critique of the US's complex history in South Korea and following the trail of toxic dumping back to Western industrial methods and Korean economic ambitions. Something monstrous emerges from the Han River in Seoul. But it is not a product of that place alone. As the creature's name suggests, it is a biological host, a carrier of toxins introduced by the US military bases in Korea and by unscrupulous industrial players, which Korea attracted by removing environmental safety restrictions.

Violations of environmental safeguards punctuate Mitchell Lichtenstein's film, *Teeth* (2007), which Roland Finger examines in Chapter 6, entitled "Biting Back: America, Nature and Feminism in *Teeth*." The small-town setting of this teen romance-thriller appears conventional in every way, with the exception of the grotesquely large smokestacks towering over the quiet neighbourhood. If this ominous image of nuclear power operates in tension with our expectations of this film genre, the instigating trauma of the story, which drives the ensuing plot, takes us even further from the type of conflict—and antagonist—we might anticipate. The central figure, high school senior Dawn, never faces off against a stock villain or a fearsome creature mutated by the town's nuclear waste sites. Rather, Dawn emerges as both the heroine *and* the menacing antagonist once she discovers that she has herself undergone a genetic mutation and developed a *vagina dentata*. With it, Dawn fends off abusive sexual advances from her abhorrent step-brother and even the local gynaecologist, whilst her affection for her class-mate, Tobey, tests the limits of her chastity vows. Innovative as it is, *Teeth* builds upon the Romantic tradition of associating women with nature, an association that has taken unique form in American culture, according to Finger. From the earliest European explorers and settlers, to the flower children of the 1970s, Americans have envisioned their country as feminine in a problematic, traditional sense: fertile, nurturing and malleable when faced with the will of determined men. Dawn embodies none of these traits, despite being a conquest of sorts for various local chauvinists. Like the landscape, violated by phallic nuclear smokestacks, Dawn is victimized ecologically and sexually at once. An uneasy narrative of vengeance unfolds that adds a traumatic edge to the film's eco-feminist politics.

In Chapter 7, Georgiana Banita similarly examines gender and violence, but takes readers into very different terrain in her analysis of Belgian director Fabrice du Welz's survival horror film, *Vinyan* (2008). Traumatized by the disappearance of their son in the Indian Ocean tsunami of 2004, a vacationing European couple faces psychological destruction when they refuse to accept the loss as a blameless tragedy. Instead they opt to venture deep into Thailand's river system in search of a mythical village of lost children. The film's heroine, Jeanne, is the driving force behind this dangerous foray, which takes the couple into claustrophobic webs of jungle that only the isolated, seemingly survivalist inhabitants know how to navigate. Motivated by Jeanne's obsessive pursuit to reclaim her (likely deceased) child, *Vinyan* brings to mind the narrative formation of the *unending quest*, derived from the epic. Deployed in countless films and other narrative forms, this type of quest has notably taken shape in George Sluizer's *Spoorloos* (aka *The Vanishing* 1988), in which a husband obsessively pursues his lost wife years after her abduction from a highway rest stop, and in Christopher Nolan's *Memento* (2000), whose protagonist seeks vengeance against the men who murdered his wife and left him with short-term memory loss. Both films present the journey as futile, but their unhinged protagonists derive their life's purpose from the quest itself. In *Vinyan*, as in these noted films, it is an initial trauma that motivates the quest—for answers, for the lost child—but it is also this lingering trauma that ensures the *failure* of the quest, or its interminability. The protagonists, shocked by the rupture, become obsessive in their quests, repetitive and redundant in their attempts to solve the mystery (literally in *Memento*) and blinded by vengeance or simply their selfish convictions that their suspicions must be correct. The maternal instinct looms large in *Vinyan*, of course, as Jeanne's childlessness motivates her descent into masochism and savagery. Jeanne and her husband, Paul, traumatized first by the unexpected power and indifference of Mother Nature, are further marked by their vengeful engagement (and climactic fusion) with nature—a formidable, traumatic ecosystem that in this context humiliates their anthropocentric worldview.

Ecological traumas such as the Asian tsunami are saddening and taxing to consider in an extended way, as Georgiana Banita does. In light of the many crises the authors explore in this book, Alexa Weik von Mossner's excavation of Andrew Stanton's animated blockbuster *WALL-E* (2008) in Chapter 8 should be most welcome, focusing as it does on *comic relief*. Disney-Pixar's robot adventure is both philosophically deeper and thematically darker than viewers expected from such populist entertainment. And humour is a key vehicle for the film's condemnation of grim postmodern realities, such as hyper-consumption and alienation. The film opens in a world that has undergone an ecological catastrophe and *not* survived to reforest itself "after people." The indicators of the disaster are everywhere in the *mise-en-scène:* consumer items, fast food wrappers, unrecyclable containers and obsolete electronics are everywhere, presumably left behind by thoughtless humans. All of this waste is stacked by WALL-E, a robot trash compactor, into towers taller than skyscrapers. And ubiquitous locations of a big box department store chain, Buy-n-Large, modelled on Walmart, lie in ruins, having fuelled human greed and supplied unhealthy appetites for more, bigger and cheaper consumer items. No humans populate the garbage city WALL-E fastidiously shapes. And director Stanton and virtuoso *Star Wars* sound designer Ben Burtt (who voices the robot) offer viewers no dialogue for the first twenty-two minutes of the film; and even then, *robots* engage in an exchange of modulated sounds. It is a brilliant irony arranged by the film that in this wasteland, devoid of humanity, a robot becomes the most humane presence. WALL-E watches Hollywood musicals in the storage locker where "he" recharges at night (more on gender ahead), learning how to desire and longing for another robot's hand to hold. EVE, an advanced version of WALL-E, arrives on earth to collect ecological samples and in the process becomes WALL-E's comedic partner as the two post-human protagonists determine whether their human counterparts have a future on earth. Disney has a long history of representing nature in peril and characters with an environmental consciousness (Murray and Heumann 2011). And yet the ecologically aware *context* in which *WALL-E* found a massive audience was surely anticipated and shaped by the prominent environmental documentaries released since the beginning of the millennium. These films have helped North Americans in particular reexamine our relationships with shopping, fossil fuels, our ecological footprint and, of course, our food.

In Chapter 9, Janet Walker examines *The Cove* (2009) as a documentary film that "eavesdrops" as a means of familiarizing audiences with dolphins in their aquatic environment, participating in pro-cetacean activism and pressing for an end to dolphin hunting, whatever the avowed purpose. The film seeks to expose and stop the Taiji, Japan, dolphin drive, in which 23,000 dolphins annually are herded into a cove and slaughtered, by deploying cameras and sound recorders in the air, on land and under water to capture the images and sounds of killing, suffering and dying. Walker explores how the film approaches the physical and epistemological challenges presented by the dolphin slaughter. By extending the usual practices of observational documentary and deploying two predominant documentary modes— "interactive" and "reflexive" (Nichols 1992)—in its diverse domain, *The Cove* permeates the borders between human and non-human species.

This intervention is ingeniously orchestrated by the film-makers and their technicians, who install cameras underwater, on the beach, on small blimps and even hidden from view in artificial rocks. Walker suggests that *The Cove* repurposes eavesdropping as a documentary mode. As a methodology for animal behaviour research that builds upon the "observational" mode of documentary, chosen because it is discrete and noninvasive, eavesdropping as mode of documentary practice is distinct because various small cameras serve as "flies on the wall," which can transmit footage of human and mammalian activity. Scrutinizing this technological and observational arrangement, Walker examines the philosophical, political and spatial variables of this trans-species experiment. The author prompts us to question which biological groups have become subject to violence and devalued as bodies. Excluded from political representation and unable to speak for themselves (a problem which *The Cove* attempts to remedy), dolphins and other animals exemplify Giorgio Agamben's concept of "bare life." The dolphins are mere raw material to satisfy Japanese tradition, to supply dubious food cartels and, above all, to service the "captivity industry" epitomized by SeaWorld water parks. What appears to be a discreet, isolated cultural practice in rural Japan reveals itself to be merely one traumatic node in a massive, distributed network. In the secluded cove, the film-makers find evidence that Japanese and US dolphin merchants, Scandinavian whaling lobby groups and national seafood policymakers represent a typical global network in the age of neoliberal globalization—one in which no single player can understand the wider structure, let alone exert agency amid its mechanisms.

In Chapter 10, "Cooling the Geopolitical to Warm the Ecological," Christopher Justice examines geopolitical tensions of a different variety, by turning to narrative films set in the Arctic, such as *The Thing* (1982) and *30 Days of Night* (2007), and providing a welcome engagement with history by surveying a range of narrative eco-trauma films produced since the 1950s. Specifically, Justice argues that scholars have overlooked the role polar ecology has played in the horror genre, and its subgenre—the creature feature. During the fifties, a decade of intense Cold War anxiety, cinematic polar landscapes took shape as menacing settings that were themselves fearful, unfamiliar and isolating. Horror and suspense were intensified, paradoxically, by the fact that the heroes were *not* entirely isolated in these barren polar landscapes, but instead subject to the hermeneutic gaze of an alien Other (*The Thing from Another World* 1951; *The Land Unknown* 1957). These narratives were, of course, allegorical representations of geopolitical power struggles and fears of invasion, takeover and even indoctrination. With their unknowable landscapes and antagonists, polar horror films were ideal vehicles for film-makers and popular audiences to engage their Cold War anxieties and the xenophobia of the period. More than allegorizing the global conflict between East and West, beneath the films' fantastical narratives of adventure and discovery loomed primal anxieties of annihilation, as well as indications that only certain forms of human action might avert a nuclear (or similar) disaster. As Justice notes, these narratives perhaps unwittingly celebrate collaborations between civilians, scientists and *military* personnel who become humanity's last hope. Routinely, the films' benevolent intellectual-researcher protagonists find themselves aligned with military solutions to the threat of the Other in the Arctic—forceful strategies that under normal conditions they would ideologically oppose.

In "Toxic Media: On the Ecological Impact of Cinema," the book's concluding chapter, Sean Cubitt leads us out of these films' complex stories and toward a consideration of our own activities and their costs. Cubitt reflects fittingly on the types of human-centrism that may be hidden in plain sight, in the very realm of media production itself. As social issue documentaries and dramas continue to find audiences and generate debate, the assumption persists that film-makers examining the earth in crisis, with budgets great and small, are attempting to use film-making for good. Surely they are. But film is also the most wasteful of the arts, with the pre-digital, photochemical process generating copious toxins in order to generate each new print. As Cubitt notes, "Despite the frequent description of the information society and the knowledge economy as 'weightless' and 'immaterial,' there is a very specific weight and materiality to the environmental footprint of contemporary media." It is well known that films are mere fragments of the miles of footage shot and developed at great expense, and discarded. Location shooting demands dozens or even hundreds of crewmembers to invade photogenic (and often pristine) ecosystems, where producers and cast members rely on private planes and helicopters. Cubitt finds such a contradiction in the *Lord of the Rings* trilogy. The films have confronted unprecedented global audiences with subtle and overt messages championing non-violence, ecological awareness and a deep respect for all living things. And yet, in their novel and ambitious digital production methods, WETA and New Line Cinema amassed a formidable ecological footprint—from rearranging a swath of earth in New Zealand to form The Shire, to running countless high temperature, high density hard drives and two thousand computers simultaneously, in order to render the digital creatures and landscapes. While considering these questions, and taking a refreshingly critical stance on the conversion to digital production and distribution, Cubitt does not disband with narrative analysis. Indeed, the films he mentions—from the masterpiece *Blow Up* (1960) to the thoughtful anti-drone thriller *Eagle Eye* (2008)—dramatize the tense relations between human and natural worlds, *as well as* foregrounding the very mechanisms we use to represent our experiences visually.

This is the challenge of our age: to overcome our disavowal, to reflect on the unsavoury forces reshaping our earth and to acknowledge our role in the production of the very mechanisms responsible for the induced eco-traumas such as e-waste and carbon emissions—even if our role is simply defined by our lack of action. Eco-trauma cinema is simultaneously about action and stasis. As the following chapters illustrate, the environmental crisis may, in many ways, confront us *as a trauma*—as an overwhelming dose of the real that lodges in our minds as an absence, and refuses intelligible recall or interpretation. In this case, our responses will be stifled.

However, if documented, filmed realities, as well as dramatized and digitally rendered worlds, can serve as our cognitive maps, aiding us in our quest to make sense of our age of anxiety, we may well *avoid* this stasis, situate ourselves as political beings and embrace our responsibilities as the earth's most empowered inhabitants.

WORKS CITED

Amorok, Tina. "The Eco-Trauma and Eco-Recovery of Being." *Shift: At the Frontiers of Consciousness* 15 (June–August 2007): 28–31. Print.

American Psychiatric Association. *Diagnostic and Statistical Manual of Mental Disorders.* 4th ed. New York: American Psychiatric Association, 1994. Print.

Anderson, Alison. *Media, Culture and the Environment.* New Brunswick: Rutgers University Press, 1997. Print.

Bouse, Derek. *Wildlife Films.* Philadelphia: University of Pennsylvania Press, 2000. Print.

Brereton, Pat. *Hollywood Utopia: Ecology in Contemporary American Cinema.* London: Intellect, 2005. Print.

Buell, Lawrence. *The Environmental Imagination: Thoreau, Nature Writing, and the Formation of American Culture.* Cambridge, MA: Belknap Press, 1996. Print.

Brockington, Dan. "Powerful Environmentalisms: Conservation, Celebrity and Capitalism." *Media, Culture and Society* 30.4 (2008): 551–568. Print.

Carson, Rachel. *Silent Spring.* Boston and New York: Houghton Mifflin, 1962. Print.

Caruth, Cathy. *Trauma: Explorations in Memory.* New York: Routledge, 1995. Print.

Cater, Nancy and Stephen Foster, eds. *Spring: Environmental Disasters and Collective Trauma.* New Orleans, LA: Spring Journals and Books, 2011. Print.

Cubitt, Sean. *EcoMedia.* Amsterdam: Rodopi, 2005. Print.

Felman, Shoshanna and Dori Laub. *Testimony: Crises of Witnessing in Literature, Psychoanalysis and History.* London and New York: Routledge, 1991. Print.

Freud, Sigmund. *Beyond the Pleasure Principle.* London: Penguin Classics, 1920. Print.

Garrard, Greg. *Ecocriticism.* London and New York: Routledge, 2004. Print.

Glotfelty, Cheryll and Harold Fromm, eds. *The Ecocriticism Reader: Landmarks in Literary Ecology.* London: Routledge, 1996. Print.

Hayles, N. Katherine. "Saving the Subject: Remediation in House of Leaves." *American Literature* 74.4 (December 2002) 779–806. Print.

Heise, Ursula K. "The Hitchhiker's Guide to Ecocriticism." *PMLA* 121.2 (2006): 503–516. Print.

Heise, Ursula K. *Sense of Place and Sense of Planet: The Environmental Imagination of the Global.* Oxford and New York: Oxford University Press, 2008. Print.

Herman, Judith. *Trauma and Recovery: The Aftermath of Violence—from Domestic Abuse to Political Terror.* New York: Basic Books, 1992.

Hochman, Jhan. *Green Cultural Studies: Nature in Film, Novel, and Theory.* Moscow: University of Idaho Press, 1998. Print.

Horak, Jan-Christopher. "Wildlife Documentaries: From Classical Forms to Reality TV." *Film History* 18.4 (2006): 459–475. Print.

Ingram, David. *Green Screens: Environmentalism and Hollywood Cinema.* Exeter: University of Exeter Press, 2000. Print.

Ivakhiv, Adrian J. *Ecologies of the Moving Image: Cinema, Affect, Nature.* Kitchener-Waterloo, ON: Wilfred Laurier University Press, 2013. Print.

Jameson, Fredric. "Cognitive Mapping." *Marxism and the Interpretation of Culture.* Ed. Cary Nelson and Lawrence Grossberg. pp. 347–360. Chicago: University of Illinois Press, 1988. Print.

Jameson, Fredric. *The Political Unconscious: Narrative as a Socially Symbolic Act.* Ithaca: Cornell University Press, 1981. Print.

Jameson, Fredric. "Reification and Utopia in Mass Culture." *Social Text* 1, Winter (1979): 130–148. Print.

Kaplan, E. Ann. "Global Trauma and Public Feelings." *Consumption, Markets and Culture* 11.1 (2008): 1–19. Print.

Kerber, Jenny. *Writing in Dust: Reading the Prairie Environmentally.* Kitchener-Waterloo, ON: Wilfred Laurier University Press, 2010. Print.

MacDonald, Scott. *The Garden in the Machine: A Field Guide to Independent Films about Place.* Berkeley: University of California Press, 2001. Print.

Mannoni, Octavio. *Clefs pour l'imaginaire, ou, L'autre scene.* Paris: Ed. du Seuil, 1969. Print.

Masson, Jeffrey. *The Assault on Truth: Freud's Suppression of the Seduction Theory.* New York: Ballantine, 1984. Print.

Minster, Mark. "The Rhetoric of Ascent in *An Inconvenient Truth* and *Everything's Cool.*" *Framing the World: Explorations in Ecocriticism and Film.* Ed. Paula Willoquet-Maricondi. pp. 25–42. Charlottesville: University of Virginia Press, 2010. Print.

Mitman, Greg. *Reel Nature: America's Romance with Wildlife on Film.* Seattle: University of Washington Press, 1999. Print.

Murray, Robin L. and Joseph K. Heumann. *Ecology and Popular Film: Cinema on the Edge.* New York: State University of New York Press, 2009. Print.

Murray, Robin L. and Joseph K. Heumann. *That's All Folks? Ecocritical Readings of American Animated Features.* Lincoln: University of Nebraska Press, 2011. Print.

Narine, Anil. "Global Trauma at Home: Technology, Modernity, *Deliverance.*" *Journal of American Studies* 42.3 (2008): 449–470. Print.

Nichols, Bill. *Representing Reality: Issues and Concepts in Documentary* Bloomington: Indiana University Press, 1992. Print.

Orwell, George. *1984.* New York: Harcourt, 1948. Print.

Radstone, Susannah. "Screening Trauma: *Forrest Gump,* Film, and Memory." *Memory and Methodology.* Ed. Susannah Radstone. pp. 79–107. Oxford and New York: Berg, 2000. Print.

Rust, Stephen, Salma Monani and Sean Cubitt, eds. *Ecocinema Theory and Practice.* London and New York: Routledge AFI Film Readers, 2012. Print.

Sandler, Elana Premack. "Avatar Blues? Is 'Avatar' Contributing to Depression and Suicidality?" *Psychology Today* 13 January 2010. www.psychologytoday.com/blog/promoting-hope-preventing-suicide/201001/avatar-blues. Web. February 2014.

Sturken, Marita. *Tangled Memories: The Vietnam War, the AIDS Epidemic, and the Politics of Remembering.* Berkeley: University of California Press, 1997. Print.

Suleiman, Susan Rubin. "Judith Herman and Contemporary Trauma Theory." *Women's Studies Quarterly* 36.1–2 (2008): 276–281. Print.

Taylor, Bron, ed. Avatar *and Nature Spirituality.* Kitchener-Waterloo, ON: Wilfred Laurier University Press, 2013. Print.

Tuttle, Brad. "Poll: Consumers Care Less about the Environment Now. Happy Earth Day!" *Time* 20 April 2012. http://business.time.com/2012/04/20/. Web. November 2013.

Walker, Janet. *Trauma Cinema: Documenting Incest and the Holocaust*. Berkeley: University of California Press, 2005. Print.

Willoquet-Maricondi, Paula, ed. *Framing the World: Explorations in Ecocriticism and Film*. Charlottesville: University of Virginia Press, 2010. Print.

Žižek, Slavoj. "Ecology." *Examined Life*. Ed. Astra Taylor. pp. 155–184. New York: The New Press, 2009. Print.

Žižek, Slavoj. *Looking Awry: An Introduction to Lacan Through Popular Culture*. Cambridge, MA: The MIT Press, 1991. Print.

1 Evolution, Extinction and the Eco-Trauma Film

Darwin's Nightmare (2004) and *A Zed & Two Naughts* (1985)

Barbara Creed

The cinema of eco-trauma raises the crucial question of humanity's humanity. What is human nature? Are we an altruistic species, which is capable of living in ecological parity with other species? Or are we either too aggressive or too fearful to share the planet and its resources? Another more disturbing possibility is that humankind is drawn not to life, but to the dark side of its evolutionary history—to warfare, decay, death and extinction. Peter Greenaway's avant-garde fiction film *A Zed & Two Noughts* (1985), or *ZOO*, offers a witty but disturbing exploration of the dark and uncanny side of Darwinian theory.[1]

A Zed & Two Noughts asks us to contemplate a series of questions central to the eco-trauma film. Why is humanity drawn to myths of life and death, creation and extinction? Which kind of thinking is more conducive to ecological survival—systematic, logical thinking or fluid or even chaotic modes of thought? Why is the human species obsessed with trying to separate itself out from the animal and natural worlds? Are we now living through what Bill McKibben has disturbingly called "the end of nature"—a concept explored by the science fiction film *Silent Running* (Trumbull 1972) in which all signs of nature (forests, birds, insects, animals, fish, fruit, waterways) have disappeared from earth, but are kept alive in spaceships hovering near the rings of Saturn.

Throughout its history, the cinema has in the main, placed the human protagonist at the centre of the narrative and given him or her agency and the power to influence events whether for good or evil. As Gorgio Agamben argues, the human has always strategically produced itself as privileged, maintaining its position through the "anthropological machine" of Western science and philosophy (33–38). This is not necessarily the case in the eco-trauma film in which the fate of human protagonists renders them impotent and calls their future existence into question. In *Twelve Monkeys* (Gilliam 1995), for instance, a malignant virus is let loose on earth. Billions perish leaving nature to reclaim the earth. A handful of human survivors are forced underground to live in virus-free spaces, prisoners of their own making. *Children of Men* (Cuarón 2006) presents the last days of the human race on earth. It is the year 2027, women have become infertile and no babies have been born for eighteen years. The human species faces its own extinction, all systems of governance have collapsed and chaos rules everywhere.

This chapter will argue that the eco-trauma film sets out to examine two central problems: the possibility of the extinction of all life on the planet and humanity's anthropocentric vision of itself. It will also propose that Darwin's theory of evolution, with its focus on anti-anthropocentrism and the role of dissolution, death, extinction and renewal, offers a rich theoretical framework for understanding eco-trauma cinema and its preoccupation with the end of the world. Darwin's writings on the evolution of emotions in man and animals, in which he argues that humans and animals experience the same emotions including pain, pleasure, joy and grief, are also of direct relevance to this discussion as they have clear implications for the way in which we treat animals.

DARWIN'S NIGHTMARE (2004)

Hubert Sauper's documentary *Darwin's Nightmare*, is aptly named. It presents a traumatic vision of the way in which human interference with a natural ecosystem not only disturbs the workings of evolution but also creates a chain of destruction which impacts negatively on the natural life and death cycle of a variety of species, including the human animal. Sauper's film examines a specific form of eco-trauma, that which is initiated by so-called human enterprise.

Charles Darwin's landmark work of 1859, *On the Origin of Species*, is central to a contemporary understanding of eco-trauma films such as *Darwin's Nightmare*. Put simply, Darwin's theory of evolution held that all living organisms have evolved very gradually over vast expanses of time from a common ancestor shared by human beings and the ape.[2] Evolutionary change is slow, and in Darwin's words "the struggle for existence" will favour some individuals with certain genetic variations over others. He said, "This preservation of favourable variations and the rejection of injurious variations, I call Natural Selection" (601). Natural selection thus gives a species a functional advantage that assists in its survival. Useful variations are preserved.

The eco-trauma film explores the different ways in which humanity's functional advantages over other species now threaten its own survival and that of the planet. In *Darwin's Nightmare* human power to introduce foreign species to new geographical areas results in an ecological disaster. The apocalyptic film *On the Beach* (Kramer 1959) reveals how man's scientific genius, once seen as a functional advantage, has resulted in nuclear war; the people of Australia, who are the last survivors, know they face extinction within months. Set in the twenty-third century, *Logan's Run* (Anderson 1976) depicts the survivors of human progress, similarly once considered an advantage, living an idyllic life in a domed city. They are kept safe from the effects of pollution and overpopulation, which, as a result of unchecked progress, are devastating the earth—the only problem is that life in Paradise ends at thirty.

Darwin's theory challenged the dominant Victorian belief that humanity came into being in a special act of divine creation, thus suggesting parity between species and creatures—an issue that is central to a number of eco-trauma films such as *Gorillas in the Mist* (Apted 1988), *Greystoke* (Hudson 1984) and *Fern Gully: The Last Rainforest* (Kroyer 1992). A major consequence of this challenge was that it presented an anti-anthropocentric critique of human values and behaviour. Gillian Beer writes that "Darwin was to rejoice in the overturning of the anthropocentric view of the universe" (32). In her book *Beasts of the Modern Imagination*, which explores representations of the animal in the Western imaginary, Margot Norris draws particular attention to the anti-anthropocentric nature of Darwin's work. She argues that Darwin's conclusions "subordinated him—the human being, the rational man—to the very Nature he studied."[3] The Darwinian revolution resulted in

> a subversive interrogation of the anthropocentric premises of western philosophy and art, and the invention of artistic and philosophical strategies that would allow the animal, the unconscious, the instincts, the body, to speak again in their work.
>
> (5)

Norris examines what she describes as the "biocentric tradition" in the work of a number of writers and artists such as Kafka, Nietzsche and Ernst. She argues that Darwinian theory abolished "the fiction of the subject as the origin of the text" (26), a position endorsed by eco-trauma films, such as *Darwin's Nightmare*, that explore themes of dissolution and extinction. The cult classic *The Planet of the Apes* (Schaffner 1968) presents a post-apocalyptic world in which human and ape have changed places, whereas the documentary *Life after People* (de Vries 2008) explores how the planet may appear if the human species disappeared. Both films abolish the human subject as the originating source of the narrative.

As Beer explains, Darwin emphasized the crucial importance of the interconnectedness of all life forms.

> Let it be borne in mind how infinitely complex and close-fitting are the mutual relations of all organic beings to each other and their physical conditions of life.
>
> (xxi)

Nevertheless Darwin was well aware of humanity's potential to upset these "mutual relations," which he saw as crucial to this theory of evolution and natural selection. In discussing artificial selection he wrote, "Man selects only for his own good; Nature only for that of the being which she tends" (qtd. in Beer xxii). Gillian Beer was one of the first to point out the ecological relevance of Darwin's theory for the contemporary world. She also noted that, as in his own day, Darwinism continues to inspire quite opposite interpretations and arguments:

> Indeed, it is remarkable that the chapters on "Natural Selection" that among Darwin's contemporaries was often read as an argument for competition "in the great and complex battle of life" now reads so strongly also as an ecological text.
>
> (xxi)

Darwin's Nightmare directly confronts the issue of eco-trauma in a Darwinian context. Directed by Hubert Sauper, *Darwin's Nightmare* examines the social, economic and environmental impact of the fishing industry on the area of Lake Victoria in Tanzania. Sauper explores the far-reaching consequences of the introduction of the Nile perch to Lake Victoria around forty to fifty years previously. The aim was to repopulate the lake, and to harvest the fish for Western markets, but the Nile perch, which is more than six feet in length, preyed so successfully on smaller fish that virtually all local species suffered extinction and the lake's ecosystem was largely destroyed. This had a disastrous effect on the lives of the impoverished and malnourished local population, who could no longer fish in the lake and were forced to work in the Nile perch industry. Unable to buy the fish for themselves, the Tanzanians are forced to live off the putrefying remains of the Nile perch, as well as the treated remains that are re-packaged to sell to them. The far-reaching destructive effects of the industry do not stop here. Sauper documents how the Russian pilots who fly the fish out also ferry in weapons to fuel the bloody local conflicts in Africa, which claim many lives. Prostitutes move into the area. As farmers leave the land to work in the new industry, food shortages create more havoc. HIV/AIDS besets the fishing community. Homeless children melt the plastic packaging to sniff the glue and become high on the toxic fumes.

As Dennis Lim points out in an article aptly entitled "The Descent of Man," "The ruthless supremacy of the Nile perch and its devastating effect on the lake's ecosystem constitute a gruesomely resonant metaphor for the impact of global capitalism on local industry." The Nile perch itself has been described by the World Conservation union as "one of the planet's 'worst invasive alien species'" (2). This cycle of events not only represents a metaphor for global capitalism, as Lim observes, but also an allegory for the narrative of human evolution in which the predatory Nile perch could be seen to stand in for the human species itself. *Darwin's Nightmare* suggests that humankind is the most "invasive" species of all. In a contest between a dominant species and those less well adapted, the giant Nile perch is the victor, just as the human appears to have triumphed over all other species. In destroying other species the human species runs the risk of destroying biodiversity on the planet and facing its own extinction. In order to survive, the Nile perch feeds on its own offspring. Eco-trauma films such as *Soylent Green* (Fleischer 1973) and *Delicatessen* (Jeunet 1991) explore post-apocalyptic themes of overpopulation, scarcity of food and cannibalism.

It is tempting to analyse *Darwin's Nightmare* in Spenserian terms as a survival-of-the-fittest narrative (a form of social Darwinism) with a focus on struggle, hierarchy, success and closure.

> Social Darwinism . . . is the spurious claim that Darwinian competition in nature constitutes a proper model for the "survival of the fittest" in human society—in which everyone competes to survive, but only the wealthy have proven themselves "fit."
>
> (Appleman 10)

This view would also endorse a "survival of the fittest" model in what has become a competition between the human species and the natural world. Those who support a free-market economy would argue that there should be no interference in this struggle. The so-called march of progress and the dominance of human needs over those of the natural and animal worlds is simply evidence that the fit will survive. Darwin, however, did not argue for the Spenserian concept of "survival of the fittest," which could be used to justify the decision to introduce the Nile Perch to Lake Victoria and its subsequent decimation of other species. Darwin stated simply that survival is a struggle. As Gillian Beer has observed, Darwin's evolutionary ideas focused on a set of concepts and values, which led to the formulation of a narrative that is very different from the Spenserian survival of the fittest storyline.

Beer sees evolutionary narrative as distinguished by a series of factors, which arise from the natural life and death cycle of all species, and which travel in opposite directions: cooperation and competition, stasis and transformation, degradation and regeneration, scarcity and abundance, continuation and extinction: "So although Darwin himself gave some considerable emphasis to the language of progress and improvement, generating an onward and upward motion in his storytelling, these tales were constantly under the pressure of other darker stories—of rapine, degradation, and loss" (Beer xix). Beer cites Darwin's discussion of the cycle of life and death: "We behold the face of nature bright with gladness, we often see superabundance of food; we do not see, or we forget that the birds which are idly singing round us mostly live on insects or seeds, and are thus constantly destroying life" (xix).

Darwin's representation of nature, however, should not be viewed as setting up oppositional sets of relationships. Rather it emphasizes flux, loss and chance and the power of inter-connecting relationships. Darwin did not

endorse the idea of "a chain of being or ladder, with its hierarchical order-
ing of rungs" but rather supported the concept of what he described as an
"inextricable web of affinities":

> These affinities he perceives sometimes as kinship networks, sometimes
> as tree, sometimes as coral, but never as a single ascent with man mak-
> ing his way upward.
>
> (Beer 19)

Critics have tended to focus on the metaphoric nature of Darwin's use
of the tree of life. Michael Mikulak persuasively argues that "the environmental crisis is more than a problem for scientists; it is a problem of narrative, ontology, and epistemology . . . we must first change the way we think
about nature if we are to solve the problem" (n.p.). He discusses Darwin's
"rhizomatic tree of life" and the writings of other rhizomatic theorists such
as Deleuze and Guattari. Mikulak is concerned about what he sees as the
limitations of rhizomatic theory, specifically its propensity to be "shaped by
other rather arborescent discourses, namely the bioscientific narratives of
biotechnology and capitalism . . ." (n.p.). Instead, he proposes the concept
of "kinship imaginaries." While I agree with much of Mikulak's argument,
I do not agree that "rhizomatic" theory offers the most accurate way in
which to understand Darwinian theory and its contribution to narrative.
As Beer demonstrates, Darwin did not draw on the symbolism of the tree
of life alone; he theorised the connectedness of all forms of life in relation
to various entities—trees, coral and kinship networks. The Darwinian "web
of affinities," with its emphasis on the "entangled" and uncanny nature of
these relationships, offers a complex and creative strategy for exploring the
themes and narrative structures of contemporary eco films.

Darwin's Nightmare offers an exemplary text for understanding Darwin's "web of affinities" and the fragile nature of such an intricate pattern
of relationships. The introduction of the Nile perch not only destroys the
lake's ecosystem but also the various other ecosystems which coexist with
those of the lake. When the people who once fished the lake are no longer
able to do so, their own future survival is threatened. When food shortages follow and the local Tanzanians are forced to move away to look for
work, prostitutes who are also looking for employment move into the area,
bringing HIV/AIDS with them. One disturbance leads to another equally
tragic event. In particular the prostitutes hope to make money from the
Russian pilots who come into the area to fly out the meat to the European markets. *Darwin's Nightmare* documents the fact that while Europeans consume thousands of Nile perch filets daily, the Tanzanians starve.
In a Spenserian sense, the stronger "fish" on land and water destroy the
weaker. However, once the web of local affinities is irretrievably broken,
and the evolutionary pattern disturbed—whether by chance, or in this
instance, opportunism—the result for the local populations (human and

non-human) is disastrous. The eco-trauma film sets out to expose the ways in which humanity, through its so-called superior knowledge and power, is able to upset what Darwin described as "mutual relations" between systems. Some may argue that "the fittest survive" but in what sense are we deploying the term "fittest"? It is possible to argue that the European entrepreneurs who have created this imbalance are technologically more advanced, but are they ethically superior? If Darwin had experienced a recurring nightmare it would be humanity's potential and power to upset the "mutual relations" necessary for evolution to flourish according the laws of nature.

A ZED & TWO NOUGHTS (1985)

More than any of Greenaway's films, *A Zed & Two Noughts* is concerned with the relationship of the human world to the animal. Greenaway draws on this relationship to explore the dark side of Darwinian theory—decay, decomposition and death as it affects all species. *A Zed & Two Noughts* explores a number of themes central to the eco-trauma film: man's persistent attempts to control the natural world through the imposition of science and systems such as taxonomies, lists and grids; man's fear of imperfection and death; and man's (as distinct from woman's) hostile relationship with the animal world. Greenaway sees evolutionary theory as liberating:

> Darwin has given us a freedom that no social or religious programme has ever given us, for, if man is on his own, then all the checks we relied on to excuse or explain our own shortcomings and mediocrities have been removed. We are, at least, now free for what we want to be.
>
> (qtd. in Pascoe 100)

Structurally, Greenaway creates in his film a series of interrelated narratives, symbols, motifs, verbal associations, puns and taxonomical associations. Films that focus on themes of eco-crisis frequently create structures that spin off each other in a complex web of Darwinian affinities. Where *Darwin's Nightmare* explored this in relation to the operations of human greed in the real world, Greenaway creates a surreal world in which the interplay of events seems endless and strangely absurd but nonetheless as similarly destructive.

The first sequence is set against the background of a giant blue neon sign—"ZOO." It is a rainy night. Two children pull a Dalmatian dog, by its leash, down the wet street. The film cuts to another scene in which a tiger paces behind the bars of its cage, roaring loudly. A severed zebra head, visible through the legs of the tiger, lies on the floor of the cage. Next we see close up of a hand, holding a stopwatch. A man sits outside the cage, timing the number of lengths of the cage that the tiger paces. As the man looks up we hear the sounds of a car crash and a woman's scream. The film cuts to the accident. A woman's head is pushed through the windscreen. The victim, whose name might be Leda, is screaming. She has a mass of red curly hair. The body of a large white bird, which turns out to be a female swan, covers the left-hand side of the screen. The camera moves in to reveal two other women lying motionless in the back seat. The swan has collided with the car. The accident occurs in Swan's Way. In this postmodern version of "Leda and the Swan," the mythical combination now proves fatal to the creature. A policeman's voice is heard: "Leda? Who is Leda? . . . Lay-da? Laid by whom?—Jupiter." A giant ESSO billboard, with the image of a leaping tiger, fills the background (Figure 1.1). In Greenaway's film, the advertising industry captures images of animals in the wild to sell products: Similarly, Sauper shows how European capitalism destroys an ecosystem to increase its wealth.

Another man, who looks very much like the first, films the movements of a one-legged gorilla, which uncannily foreshadows future images of the one-legged, red-haired woman who survives the accident. The gorilla approaches the man. Its image is replaced by a close-up shot of the camera and its single frame mechanism. The early history of the cinema developed from the stop-motion photographic experiments of Eadweard Muybridge, who used this technology in his pioneering studies on animal movement. Thus, the history of the cinema is closely associated with the image of the animal—as Greenaway reminds us.

Figure 1.1   The Esso logo and the oil company's mascot, the tiger, are revealed in *A Zed & Two Naughts*.

In an uncanny revelation, we learn that the two dead women in the car are the wives of two doctors, identical twins, named Oliver and Oswald Deuce. Images of doubling, twins and freak accidents invoke a sense of the uncanny, of the familiar becoming unfamiliar. Unable to assuage their grief, the brothers first turn to science for comfort—and answers. Their faith in science initiates their growing fascination with evolution, decomposition and death. "I cannot stand the idea of her rotting away," says Oswald, who rails against the natural process of evolutionary decay. "How fast does a woman decompose?" Oliver replies. "Six months, maybe a year." Oswald asks what happens when a body decomposes. Oliver explains, without any display of emotion. "The first thing that happens is that bacteria set to work in the intestine. . . ." He tells Oswald that there are 130,000 bacteria "in each lick of the human tongue . . . the first exchange was at the very beginning of creation when Adam kissed Eve."

The next scene is of the twins throwing flowers on the ground at the crash site. Their mourning is disturbed by the sounds of David Attenborough's voice-over in an extract taken from his *Life on Earth* (1979) series. Drawing on Darwinian evolutionary theory, Attenborough gives a very different version of the beginning of life.

> If the evolutionist span of life on Earth is represented by a year of 365 days then man made his first appearance on the evening of thirty-first of December just as the daylight was breaking. It had taken some 4000 million years for that entrance to be made. A very slow and uneven progress of life forms, changing and evolving in a long continuous procession.
>
> (n.p.)

As Attenborough discusses man's evolutionary history, the screen fills with a large close-up of a snail on a brightly coloured pink flower—a creature, like Darwin's worms, that feeds on decomposing matter, reminding us that death is the fate of all living things. Attenborough explains that it is comparatively easy to understand how one species gave way to another, but more difficult to understand "the evolutionary leap necessary to bridge the most sophisticated of apes with man and more difficult still to comprehend how life could create itself apparently out of nothing." The conditions for the creation of life on earth, the appearance of blue-green algae, can never be reproduced again. (Similarly, nor will the evolutionary conditions necessary for Lake Victoria to rebuild its ecosystems ever reoccur in the right sequence. This is another face of evolution's nightmare in Sauper's film). Chance gave rise to human life—just as chance led to the ill-fated car crash outside the zoo in which human and animal become strangely entangled. Entanglement, which is central to Darwin's theory of evolution, is represented by Greenaway as antithetical man's concept of civilization. A central theme of the eco-trauma film is the creative power of entanglement.

At relevant moments, Greenaway intersperses sequences from David Attenborough's documentary throughout the film in order to contrast Darwin's view of the origins of species with other explanations—biblical and mythical. The Deuce brothers set out to understand their grief through a scientific study of evolution. Oliver concentrates on the Attenborough documentary, asserting that he finds it is cathartic to watch "the beginnings of life." Oswald spends time observing signs of death as he films various dead animals from the zoo in stages of decay.

The first sequence on the beginning of life comes to an abrupt end as the image of Leda, lying on a hospital bed, fills the screen. Reference to the myth of "Leda and the Swan" offers a classical version of the beginnings of life—new forms emerged from the coupling of human and animal. It is the Darwinian version, however, that dominates. Gradually the two brothers turn to Leda, a Frenchwoman whose real name is Alba Bewick, for emotional and sexual support, as well as answers to their questions. Her daughter, Alpha, is learning the letters of the alphabet by associating each letter with an animal—suggesting a basic connection between animals and language. Alba, who has lost a leg in the accident, is confined to her bed, eventually deciding that in the interests of symmetry she will have the other leg amputated. Scenes from the Attenborough documentary attest to the importance of symmetry in nature. "Why do we have to have two of everything?" Alba asks. "Symmetry is all," Oswald answers, leading us to recall the importance of symmetry in the biblical stories of the twins, Cain and Abel, and the symmetry behind the pairing of animals in the story of Noah's ark. She eventually bears the Deuce twins, a set of twin offspring. The film's uncanny sets of doubles continue to multiply. Symmetry imposes its own order, but threatens to create an uncanny disorder when taken to extremes.

Alba tells Oliver she first met his wife at the butterfly enclosure: "Your wife said they should be let free. She didn't approve of zoos." Shortly after, Oliver frees the butterflies from their cages. Oswald tells Alba also he can't stand the idea of his wife rotting away. He picks up a photograph of an idyllic country scene and enquires about its location. Alba tells him it is where she was born—L'Escargot. Oswald wants to know exactly what his wife did before her death, and what she bought when shopping. Alba tells him she bought prawns. The scene cuts to a close-up of a prawn, its small legs moving rapidly as it swims through the water. Oliver is watching the sequence from *The Beginnings of Life* on crustaceans. Attenborough explains that this is an "extraordinary" period—"side-scuttling crabs, slow-moving snails . . . animals that live on detritus and each other, creating systems of defence and attack, whose ingenuity is manifest." In contrast to the crustaceans, Oliver's life is becoming less complicated. He has stopped bathing and is beginning to smell. Oswald films the slow decomposition of prawns. They are "on their way back from where they came from—ooze, slime and murk." Oswald screens an episode of the Attenborough documentary to Alba. He tells her he is trying to work out the mystery of evolution. "Why we should come all this way, slowly and painfully. Inch by inch. Fraction by fraction. Second by second—so that my wife should die by a swan?" Without success he continues to search for a meaning to his wife's freak accident.

The other major female character is a prostitute and seamstress called Venus de Milo (the famous statue, of course, has no arms), whose trade name is Venus, but her clients call her Milo. She empathises with the animals in the zoo and also writes dirty stories. One of her customers, Van Hoyten, tries to exchange sex for animal meat (zebra steak, calves liver), while implying that Milo has sex with animals in the zoo. In contrast to Oliver, Oswald and Alba, Milo is not interested in maintaining symmetries. She represents a strange and seductive world of sex, storytelling, nursery rhymes and animals. After having sex with Milo, Oliver asks her if she has "ever done it with animals." If that is what he wants, she will oblige. She could also tell him stories, in the manner of Anaïs Nin. Sitting up in bed, Oliver arranges snails on a glass plate—several crawl across his body and he places one in his navel. Plants and ferns fill the background. The operations of chance, central to evolutionary theory, now dominate. His familiar, ordered scientific world is breaking down and becoming unfamiliar, assuming an uncanny dimension. The key definitional aspect of the uncanny is where the familiar becomes unfamiliar—or strange. Venus tells him a story about a polar bear called Fairbanks, who had a "narrow snout," a "sweet nature" and a "rough and probing tongue" and who was kept to entertain women. Oliver tells her he likes snails because they are "a nice primitive form of life. They help the world decay." They are also hermaphrodites and can "satisfy their own sexual needs." When Milo tells Oliver she disapproves of zoos, he throws her out. In an eerie scene, she laughs and walks down the stairs, donning her clothes and singing "Teddy Bears' Picnic." Later she tells Oswald a story about a toad ("at least it had the body of a toad") that was kept in the Regents Park Zoo in London in the 1870s in the Obscene Animals Enclosure, which was reserved for certain rare animals.

In her discussion of *A Zed & Two Noughts*, Amy Lawrence draws attention to the way in which Milo collapses boundaries between species: "The children's nightmare/fantasy of animals taking over when it is dark and no one is looking coincides with Milo's fantasy of a world of animals freed from human dominance" (82). Whenever the song is played in the film, it invokes a sense of the uncanny—of teddy bears coming to life at night and engaging in erotic games. The familiar surfaces are displaced by strange events. Rather than try to control the world through the imposition of rationality and order, systems and taxonomies, Venus de Milo values sex, the power of the imagination and the animal world. Like Oliver's wife, she does not like zoos. Gradually Oliver and Oswald begin to change their daily lives. Through their relationship with Milo, and later Alba, they are each brought more and more into the world of nature. They are drawn to snails, decaying matter, the dark woods of the "Teddy Bears' Picnic," and the countryside of L'Escargot, Alba's country birthplace. The zoo attendants huddle around

Oswald's bowl of prawns (he is filming their decomposition); Van Hoyten wonders if he is trying to recapture the smells of his wife. Oliver throws a bowl of flowers over Alba, telling her that he knows she was pregnant when she was driving the car that killed his wife and that "pregnant women are notoriously unreliable." He also says that she was wearing white feathers, which must have attracted the swan. "You were asking for trouble!" Under Milo's influence, Oliver is abandoning his normally scientific and rational approach to life which he once used to control the forces of nature, decay and death.

Greenaway uses repeated images of the zebra to introduce a discussion of Darwin's theory of sexual display and sexual selection (Figure 1.2). Milo and the legless Felipe Arc-en-ciel are in one of the cages, directly in front of the zebra, which is running back and forth behind the bars. Felipe says to Milo, "If the zebra is such a beautiful animal you'd have thought that men would have invented a fanciful hybrid. Wouldn't you? . . . Half woman. Half zebra. . . . Ever ready haunches." "They'd only put it in a zoo," says Milo. Felipe muses, "Animals are always kept for profit. . . . If I had the money to own a zoo, I would stock it with mythological animals." The film cuts to a tight close-up of a pattern of black and white stripes; the camera pulls back to reveal that the stripes forms the pattern of a coat worn by an artist, who is painting a naked woman, wearing a red hat (from Vermeer's painting *The Girl with the Red Hat*) and holds a trumpet. The scene presents an uncanny re-enactment of Vermeer's famous work *The Art of Painting* (1673) in which Vermeer is the artist. There are further references to black-and-white zebra patterns on the dresses of Vermeer's female subjects in *The Music Lesson* (1664) and *The Concert* (1665). Milo is commissioned to make the same zebra dress for Alba, who wears it as she sits playing the piano in imitation of the woman in *The Music Lesson*. Three key differences disturb the otherwise tranquil scene and render the familiar unfamiliar: Alba has a prosthetic leg; Vermeer's dress is stitched, along with Alba, onto the piano stool; and Alba beats out "Teddy Bears' Picnic" on the piano. In a later scene, Alba lies asleep as terrible screams from the zoo rent the night air. The next day the zebra is found dead. Later the twins film the zebra and its gradual decomposition. Greenaway is creating a surreal world dominated by evolutionary themes of chance and the close relationships of human/animal.

Throughout *A Zed & Two Noughts*, Greenaway constantly draws uncanny associations between women and animals. The gorilla and Alba have each lost a leg. Alba has her other leg amputated, and Van Meegeren[4] plots to remove the gorilla's remaining leg. Alba, who wears the zebra dress, plays "Teddy Bears' Picnic" on the piano and a recording of "The Elephant Never Forgets" on her gramophone. She lives in an apartment close to the zoo—at night she hears animals' cries. She doesn't approve of zoos and tells Oswald that his wife felt the same way. Milo seems to live in the zoo. She meets her customers there and several refer to having sex with Milo

Figure 1.2   The zebra appears on screen in *A Zed & Two Naughts*.

in the animal enclosures. Milo entertains the twins and other customers with erotic stories about bestiality. She wants to set the animals free. The wives of the Deuce twins die in an accident caused by a swan. The twins cannot bear to think of their decomposing bodies, yet commence a series of experiments to record on film the decomposition of dead animals from the zoo. Van Meegeren tells Oswald he has noticed that all of his experimental subjects have been female animals. The twins, on the other hand, use their photographic experiments to erect a barrier between themselves and the world of the zoo animals. By constructing the dead and decaying animals as objects of scientific experiments, they attempt to control and regulate the animal world.

In general, critics have not discussed in depth Greenaway's representation of the animal world in relation to ecological considerations. *A Zed & Two Naughts* is seen more as an example of Greenaway's love of form and symmetry, Dutch painting, Vermeer, taxonomies and theoretical puzzles. However, questions of eco-trauma are central. Greenaway has explained that he wanted to examine "the trauma of loss and the fascination of decay" and to offer "a platform for consideration, without judgment, of man's persistently dubious relationship with animals . . ." (Espejo 3). Greenaway investigates humankind's propensity to control the natural world through systems—zoological taxonomies, the alphabet, collections, cages, preservation, lists and grids. The favoured approach is stop-frame photography. The twins, who work in a zoo, set out to explore their grief through the implementation of such systems. Oswald sets about filming, frame by frame, the decomposition and decay of a variety of objects and creatures: apple, prawns, fish, crocodile, Dalmatian, zebra, the one-legged gorilla and finally themselves—the human species, supposedly the most advanced of all. As Greenaway has noted, in undertaking their study the twins have followed Darwin's stages of Evolution "until finally . . . only one experiment remains to be done: the filming of the decomposition of human bodies" (qtd. in Pascoe 7). In the end, they decide to film their own decaying bodies. The scenes of animal decomposition are filmed in the zoo's laboratory, which is filled with test tubes, glass jars and other containers, designed to confine, control and measure various species plant and animal from the natural world. By filming and recording the stages of decomposition, the twins mistakenly imagine that they will eventually learn to control their grief and exert mastery over death.

Greenaway has said that he wants "to make films that rationally represent all the world in one place . . . to pull all the world together" (Pally 6). Greenaway uses the plight of the Deuce twins to talk about man (as distinct from woman) and his desire to acquire mastery over animals, nature and death. The zoo is the grand signifier of this impossible desire. In his essay "Why Look at Animals?" John Berger discusses the significance of the human desire to capture images of animals through photography:

> In the accompanying ideology animals are always the observed. The fact that they can observe us has lost all significance. They are objects of our ever-extending knowledge. What we know about them is an index of our power, and thus an index of what separates us from them. The more we know, the further away we are.
>
> (14)

The ideology of zoos is the same. A zoo represents a brutal attempt to control nature—to classify and confine its many species, while separating them from the human. Ironically, in attempting to control and inevitably to destroy nature, humankind approaches its inevitable end with even greater rapidity—for humankind has forgotten it cannot live without nature and the complex world of nature's biodiversity. The zoo also represents an attempt to separate human and animal in order to deny Darwin's theory of a common ancestor.

Depicted as somewhat of an outcast, Milo is the only character in the film who embraces herself as animal. In her last scene, she appears at the zoo gates at night and picks the padlock. "ZOO" now reads "OOZ." Nothing is as it should be in the zoo. Milo walks barefooted through the gates and across the lawn to the sounds of horses neighing and other strange welcoming animals' cries. Milo has finally embraced the animal world—emotionally and sexually. Meanwhile Alba tells the twins that they can have her body. She dies with the twins lying symmetrically on either side of the bed. The record on the gramophone sings of animals. The ditty "An Elephant Never

Forgets" fills the room. What is it that elephants never forget? Their lives made wretched in human servitude and the zoo?

In the final scene, the twins travel to the country retreat of L'Escargot, an Edenic retreat. Alba tells Oswald, "It's full of snails and butterflies." The snails "lick up your sweat." The twins decide to film their own deaths, decay and putrefaction. It is as if they have decided in their final hour to relinquish systems that are antithetical to ecology and to embrace the forces of evolution. They have erected a stage and set up their camera equipment. There is even a gramophone, which plays "Teddy Bears' Picnic." They carefully arrange their naked bodies on a board, carefully marked by the lines of a grid, so that they become uncannily indistinguishable from the other creatures, whose decomposing bodies they also pinned to a grid and filmed with scientific precision. The twins inject each other with drugs. In the end, we see the snails, which help the processes of decay, waiting to feast upon decomposing matter. A strange hermaphroditic creature, which when young will cannibalise unhatched eggs, the snail whose tongue is covered with tiny teeth, fits in perfectly with the film's fascination with decay and decomposition. With its focus on death and the inevitability of natural processes, the final scene of *A Zed & Two Noughts* suggests the extinction of the human species. Hundreds of snails smother the camera equipment, causing a short circuit and power breakdown. The camera explodes in a flash of light. The snails continue to cover everything man-made—they are reclaiming their territory. Nature it appears has won. The twins have become the "two noughts" of the flashing neon sign—the "Z" having ceased to glow at all in some earlier scenes. The film's anti-anthropocentric sensibility ends in the irrevocable gesture of human extinction—man has "come to nothing."

David Pascoe sees the twins' attempts to understand the death of their wives through a study of evolution, decomposition and extinction as having no purpose:

> In this respect, the grief-stricken twins are themselves Muybridgean subjects for Greenaway's camera. There are no purposes only functions. The actions of the body—and its decay—are demonstrated, and every step of a movement is made look as significant or insignificant as every other step. This lends Oliver and Oswald a pathos and helplessness, as they vivisect, and then are vivisected, anatomized into temporal fragments. This is the heart of the film. . . .
>
> (111)

While Pascoe's observations are illuminating, they do not take into account the other "heart of the film"—the animal, which Greenaway has acknowledge is central. In terms of eco-trauma the heart of the film is the zoo itself—a place of imprisonment, cruelty and experimentation. The zoo signifies a major breakdown in the relationship between human and the ecology. The zoo is designed to separate the human animal from other species, to create a boundary, a chasm—to enable human to define itself as not animal. Captive, decapitated, maimed, dead, decomposing—Greenaway's animals do not exist in their own right. They exist only in relation to what they signify for the human. The absence of the animal as a subject in its own right lies at the heart of the trauma in *A Zed & Two Noughts*.

In discussing the film, Greenaway expresses concern about the way in which humanity has separated itself from nature and from its origins:

> It could be said now that all animals live in zoos, whether it is a zoo in Regent's Park, London or a Nigerian Game Reserve. Perhaps what's left to argue is only the zoo's quality. Thanks to the Voyage of the Beagle, the demise of the work-horse, the plans now being made to leave Earth and our present dubious passion with ecology, our relationship with animals has changed dramatically in the last hundred years; but has our sense of responsibility improved? And have we acknowledged another responsibility? What about mermaids, centaurs, the Sphinx, the Minotaur, werewolves, vampires and that proliferating zoo of contemporary hybrids. If one parent was an animal now familiar behind bars in the zoo, who was the other?
>
> (Espejo 3)

Greenaway's interest in mythical human and animal hybrids raises the fraught question of humanity's origins and the spectre of the uncanny. Is this the face of humankind's "other" parent? *A Zed & Two Noughts* uses Milo, the prostitute, who embraces the idea of bestiality (and a common ancestor), to keep this motif in the forefront of the narrative.

ECO-TRAUMA FILMS

A Zed & Two Noughts shares a number of ideological themes with other eco-trauma films, particularly themes about the end of nature and the extinction of species, both of which *Darwin's Nightmare* also explored. These themes develop from a Darwinian form of the uncanny (Creed 1–18). Darwin's theory of a common ancestor, shared between ape and human, disturbed all known categories concerning human origins. Post-Darwin, the human animal is forever tied to its double, the ape, and human culture to its double, nature. Throughout Greenaway's film, the appearance of doubles invokes the uncanny that is a sense of the familiar rendered unfamiliar. The characters mirror each other and the human world mirrors that of the zoo. Nicholas Royle discusses the implications of the uncanny for the natural order. "It is a crisis of the natural, touching upon everything that one might have thought was 'part of nature': one's own nature, human nature, the nature of reality and the world" (Royle 1).

Sigmund Freud, who acknowledged his indebtedness to Darwinian theory, discussed the uncanny in his classic essay of 1919, *Das Unheimliche*
("The Uncanny"). He argued that what distinguishes the uncanny is a sense
of the familiar made unfamiliar (*unheimlich*). The uncanny is related "to
what is frightening—to what arouses dread and horror" (Freud 339). In
particular the uncanny is related to loss—in particular loss of limbs, loss of
human form (in the appearance of automata), loss of sight. Here, I would
also add humanity's loss of its "natural" self—its other being that belongs
to the world of nature. What haunts the eco-trauma film is such a loss—
the possibility that nature itself may become extinct as Bill McKibben has
warned: "An idea, a relationship, can go extinct just like an animal or a
plant. The idea in this case is 'nature', the separate and wild province, the
world apart from man to which he adapted, under whose rules he was born
and died" (44–45). Eco-trauma films tap into the "strangely unfamiliar" in
that the *mise-en-scène* seems perpetually haunted by the absence of something once important that has now been lost—a central theme of *Darwin's
Nightmare*. The eco-trauma film is haunted by doubles: life and death, survival and extinction, human and animal, culture and nature. Because the
web of affinities in the ecological order has been destroyed, the image or
memory of a past sustainable world haunts the image of the present ruined
world. The uncanny of eco-cinema represents in Royle's terms "a crisis of
the natural."

We can see this dynamic at work across various forms of the eco-trauma
film. These include films about extinction; films about the collapse of the
bond between nature and culture; and films about a post-evolutionary
world of replicants and repetition. Eco-trauma films, which focus on extinction, explore the threat that humanity itself presents to the future of the
planet. This threat may come from overpopulation and lack of food and
resources (*Soylent Green*, Richard Fleischer 1973), nuclear war (*On the
Beach*, Stanley Kramer 1959 and *The Planet of the Apes*, Franklin J. Schaffner 1968) and plague (*Twelve Monkeys*, Terry Gilliam 1995). Variations on
the uncanny run through these texts, rendering the familiar unfamiliar in
relation to Darwinian motifs of evolution/devolution, abundance/scarcity,
regeneration/extinction.

Eco-trauma films that investigate the breakdown of the relationship
between the human and natural worlds similarly draw on the Darwinian uncanny. A central theme is the fragile nature of the boundary that
exists between species—the human and non-human animal. Each side of
the binary bleeds into the other, creating an uncanny dimension that is perpetually present. Agamben explores this relationship in *The Open* in which
he argues that the human subject has produced itself strategically as completely separate from the animal, creating a caesura between man and animal that raises man above the animal and nature, and at the same time
locates animality outside of human reach, outside of what Heidegger termed

humankind's once natural openness to the world. Thus, the human animal is uncanny because in its attempts to repress its animal origins, it no longer "lives in" the natural world, but appear as a stranger to itself—no longer at home in its own skin.

Some eco-trauma films explore the desire to renew bonds between the human and natural worlds as in *Tarzan the Ape Man* (Van Dyke 1932), *Greystoke: The Legend of Tarzan, Lord of the Apes* (Hudson 1984), *Beauty and the Beast* (Cocteau 1946), *Gorillas in the Mist* (Apted 1988), *Wolfen* (Wadleigh 1981), *Wolf* (Nichols 1994) and *Max Mon Amour* (Oshima 1986). The Tarzan films explore the divide between nature and culture through Tarzan's uncanny and indeterminate status as man and ape. In films such as *Wolf* and *Greystoke*, the human protagonist rejects civilization for the natural world. *Max Mon Amour*, a French satire on the bourgeoisie, depicts Charlotte Rampling in a relationship with a chimpanzee, much to the chagrin of her unfaithful hypocritical husband. The relationship between woman and animal is uncanny in an amusing context in that it becomes strangely familiar in its mirroring of her marriage—both fraught with secrecy, petty jealousies and possessiveness. Despite its humorous intent, the film makes it clear that this surreal state of affairs is largely a result of the humanity's failure to maintain any kind of productive, non-exploitative relationship with the natural world, symbolised by the fact that Rampling first encounters Max in a zoo. Peter Jackson's *King Kong* (2005) similarly explores themes of eco-trauma in its focus on the relationship between culture and nature. In a new scene, not in the 1933 original version, Kong takes Ann to his cave. Together they sit watching the sunrise. "It's beautiful!" says Ann, looking up first at the sun and then at Kong. Peter Jackson's film points to the theme of eco-trauma by mourning the end of Nature on Skull Island. Kong is captured, chained and put on display, but it is the human predator/spectator who uncannily resembles, not the civilized and superior being, but the supposedly amoral beast.

A third group of eco-trauma films explore what is best described as post-evolutionary theory. These films are about automata, cloning, replication and repetition. They include *Blade Runner* (Scott 1982), *Gattaca* (Niccol 1997) and *Alien Resurrection* (Jeunet 1997). Although the Darwinian body is essentially fluid and malleable, it is still subjected to the processes of natural selection that take place over vast periods of time. Darwin's theory emphasized the crucial importance of variation in the natural world and the possibility of monstrosities. In the eco-trauma film, the post-Darwinian human body is one that has been subject to transformation, but not in the natural evolutionary process. This body, which is a futuristic one designed for enhanced survival, is no longer a part of nature or the natural eco-system and as such subject to disease, decay and dissolution. Characters that have been genetically engineered or artificially created are familiar yet unfamiliar; they are "human" yet strangely similar. Eco-trauma films, such as *Gattaca,* engage with the possibility of post-evolution in that they create a world in which the evolutionary process is scientifically altered to remove the influence of chance events and unexpected transformations in the process of human reproduction—in allegorical terms the human has won, evolution is over and nature is dead.

In conclusion, I wish to refer to one of the bleakest moments in Darwin's writings. His description, from the Beagle diary, of the arid wastes of Patagonia is almost prescient in its account of a landscape in which nature already appears to have died.

> In calling up images of the past, I find the plains of Patagonia most frequently cross before my eyes. Yet these plains are pronounced by all most wretched & useless. They are only characterized by negative possessions;—without habitations, without water, without trees, without mountains, they support merely a few dwarf plants. Why then, and the case is not peculiar to myself, do these arid wastes take so firm possession of the memory? Why have not the still more level, greener & fertile Pampas, which are serviceable to mankind, produced an equal impression?
>
> I can scarcely analyse these feelings.—But it must be partly owing to the free scope given to the imagination. They are boundless, for they are scarcely practicable & hence unknown: they bear the stamp of having thus lasted for ages, & there appears no limit to their duration through future time.
>
> (773–774)

Darwin is unable to forget the "most wretched" plains of Patagonia. This is a post-apocalyptic landscape—a scene of ecological devastation. Perhaps this scene took such a firm hold on Darwin's imagination because deep down he knew that the human species had the power to render the earth inhospitable to all species. This is Darwin's nightmare.

NOTES

1. Greenaway has demonstrated a strong interest in Darwinian theory in an earlier film, *Darwin* (1992), made for Channel 4 in the UK.
2. Although Darwin did not say that the human species was evolved from the ape, he did say that humans and chimpanzees shared a common ancestor. The relatively recent discovery of DNA, and the mapping of the human genome, which reveals that human and chimpanzee share 98.4% of their DNA, has again sparked controversy over the origin of the human species.
3. Norris's book, *Beasts of the Modern Imagination*, represents one of the first attempts to explore Darwin's anti-anthropocentrism and the representation of the animal in Western culture. The quote is from the dust jacket.
4. Van Meegeren, the surgeon, has the same name as the infamous Vermeer forger and is said to be related. He is a sinister figure who seems to take pleasure from amputating Alba's legs.

WORKS CITED

Alien Resurrection. Dir. Jean-Pierre Jeunet. Twentieth Century Fox, 1997. Film.

Agamben, Giorgio. *The Open.* Stanford: Stanford University Press, 2004. Print.

Appleman, Philip. *Darwin.* New York: Norton & Company, 1970. Print.

Beauty and the Beast. Dir. Jean Cocteau. Lopert Pictures, 1946. Film.

Beer, Gillian. *Darwin's Plots: Evolutionary Narrative in Darwin, George Eliot and Nineteenth-Century Fiction,* Cambridge: Cambridge University Press, 2000. Print.

Berger, John. "Why Look at Animals?" in *About Looking.* London: Writer and Readers Publishing, 1980, 3–30. Print.

Blade Runner. Dir. Ridley Scott. The Ladd Company and Warner Bros, 1982. Film.

Children of Men. Dir. Alfonso Cuarón. Universal Pictures, 2006. Film.

Creed, Barbara. *Darwin's Screens: Evolutionary Aesthetics, Time and Sexual Display in the Cinema.* Melbourne: Melbourne University Press, 2009. Print.

Darwin, Charles. "Darwin's Beagle Diary" (1831–1836). Transcribed by Kees Rookmaaker from the facsimile published by Genesis Publications, 1979. Ed. John van Wyhe. *The Complete Works of Charles Darwin Online,* 2006, 1–799. http://darwin-online.org.uk. Web. Accessed 30 November 2013.

Darwin, Charles. *"The Origin of Species" and "The Voyage of the* Beagle." New York: Alfred A. Knopf, 2003. Print.

Darwin's Nightmare. Dir. Hubert Sauper. International Film Circuit, 2004. Film.

Delicatessen. Dir. Jean-Pierre Jeunet. Miramax, 1991. Film.

Espejo, Daniela. "A Zed & Two Noughts." http://petergreenaway.org.uk/zoo.htm. Web. Accessed 30 November 2013.

Fern Gully: The Last Rainforest. Dir. Bill Kroyer. Kroyer Films and Twentieth Century Fox, 1992. Film.

Freud, Sigmund. "The Uncanny." *Pelican Freud Library,* Vol. 14. Ringwood: Penguin, Australia, 1975, 335–376. Print.

Gattaca. Dir. Andrew Niccol. Jersey Films, 1997. Film.

Gorillas in the Mist. Dir. Michael Apted. Universal Pictures, 1988. Film.

Greystoke: The Legend of Tarzan, Lord of the Apes. Dir. Hugh Hudson. Warner Bros., 1984. Film.

King Kong. Dir. Peter Jackson. Wingnut Films and Universal Pictures, 2005. Film.

Lawrence, Amy. *The Films of Peter Greenaway.* Cambridge: Cambridge University Press, 1997. Print.

Life after People. Dir. James de Vries. Flight 33 Productions, 2008. Film.

Life on Earth: A Natural History by David Attenborough. Prod. Christopher Parsons. BBC Natural History Unit, 1979. Film.

Lim, Dennis. "The Descent of Man." *Village Voice,* 26 July 2005. www.villagevoice.com/2005-07-26/film/the-descent-of-man/full/. Web. July 2014.

Logan's Run. Dir. Michael Anderson. Metro-Goldwyn-Mayer, 1976. Film.

Max Mon Amour. Dir. Nagisa Oshima. Toho, 1986. Film.

McKibben, Bill. *The End of Nature.* London: Viking, 1990. Print.

Mikulak, Michael. "The Rhizomatics of Domination: From Darwin to Biotechnology." *Rhizomes* 15, Winter (2007). www.rhizomes.net/issue15/mikulak.html. Web. Accessed 30 November 2013.

Norris, Margot. *Beasts of the Modern Imagination: Darwin, Nietzsche, Kafka, Ernst, and Lawrence.* Baltimore and London: The Johns Hopkins University Press, 1985. Print.

On The Beach. Dir. Stanley Kramer. United Artists, 1959. Film.

Pally, Marcia. "Cinema as the Total Art Form: An Interview with Peter Greenaway." *Cineaste* 18.3 (1991): 6–11. Print.

Pascoe, David. *Peter Greenaway: Museums and Moving Images,* London: Reaktion Books, 1997. Print.

The Planet of the Apes. Dir. Franklin J. Schaffner. APJAC Productions and Twentieth Century Fox, 1968. Film.

Royle, Nicholas. *The Uncanny*. Manchester: Manchester University Press, 2003. Print.

Silent Running. Dir. Douglas Trumbull. Universal Pictures, 1972. Film.

Soylent Green. Dir. Richard Fleischer. Metro-Goldwyn-Mayer, 1973. Film.

Tarzan the Ape Man. Dir. W. S. Van Dyke. Metro-Goldwyn-Mayer, 1932. Film.

Twelve Monkeys. Dir. Terry Gilliam. Atlas Entertainment, 1995. Film.

Wolf. Dir. Mike Nichols. Columbia Pictures, 1994. Film.

Wolfen. Dir. Michael Wadleigh. Orion Pictures, 1981. Film.

A Zed & Two Naughts. Dir. Peter Greenaway. Artificial Eye and Skouras Pictures, 1985. Film.

2 Trauma, Truth and the Environmental Documentary

Charles Musser

During the first decade of the twenty-first century, a host of environmental issues related to global warming, energy, pollution and our food supply became increasingly urgent even as US president George W. Bush and other world leaders refused to take them seriously. Documentary film-makers responded, and by the end of the decade, the environmental documentary had emerged as the pre-eminent genre in this nonfiction mode, at least in the US and Europe. The 2009 Sundance Film Festival screened six documentaries dealing with environmental issues and was described (mockingly in some cases) as the "green festival" (Cieply n.p.). These included Joe Berlinger's *Crude* (2009), Robert Stone's *Earth Days* (2009), *Dirt! The Movie* (directed by Bill Benenson, Gene Rosow and Eleonore Daily, 2009), Rupert Murray's *The End of the Line* (2009), John Maringouin's *Big River Man* (2009) and Louie Psihoyos's *The Cove* (2009), which would go on to win the Academy Award. This outpouring of top-flight environmental films was not unique to Sundance: A few months earlier, the 2008 Toronto Film Festival had showed Dan Stone's *At the Edge of the World* (2008), about radical members of the anti-whaling movement; Ben Kempas's *Upstream Battle* (2008), about Native Americans fighting to preserve the rivers and their culture of salmon fishing; and *Food, Inc.* (2008), which was nominated for an 2009 Academy Award and was considered by many to be *The Cove*'s major Oscar competition. There was also at least one Disney documentary, *Earth*, which was playing in overseas films festivals in 2007 and 2008 but was released in the US only in April 2009—on Earth Day.

Several aspects of this formation are of concern in this chapter. The first is historical: how to situate this explosion of environmental documentaries in relationship to an earlier history and pre-history. In this regard, Bill Nichols suggestion that documentary practice as a specific formation can be understood from three vantage points is helpful—through (1) the self-understanding of its practitioners, (2) the texts that are the product of that practice and (3) a constituency of viewers (Nichols, *Representing* 17–28). These elements are also at play in the formation of this now widely accepted genre. In this respect, the organizers of environmental film festivals have played a crucial role in its constitution over the last twenty years. We might see these pioneering figures as active viewers who became members of the community of practitioners, which includes not just film-makers but critics, distributors, scholars and (of course) festival organizers. The dynamic formation of the environmental documentary has depended on the interactions between a rapidly changing documentary practice and a dynamic environmentalism, each with its own complex history.

A second area of investigation concerns the ways in which environmental documentaries are nonfiction instances of what Anil Narine calls eco-trauma cinema. In fact, these documentaries engage "the harm we, as humans, inflict upon our natural surroundings, or the injuries we sustain from nature in its unforgiving iterations." In some respects it has been easier to offer powerful instances of eco-trauma in fiction films such as *Soylent Green* (1973) or *The Day after Tomorrow* (2004). Although climate change, pollution and threats to our food supply may not yet have produced obvious traumas on a worldwide scale, these global crises often occur more locally or selectively as physical traumas to specific ecosystems. Many of the people who appear in environmental documentaries are traumatized by ecological events and devastation. Their lives have been upended, and they feel compelled to speak— to bear witness to their trauma often as a way to begin to take action and also begin the process of recover. As Judith Herman has noted,

> The core experiences of psychological trauma are disempowerment and disconnection from others. Recovery, therefore is based upon the empowerment of the survivor and the creation of new connections. Recovery can take place only within the context of relationships; it cannot occur in isolation.
>
> (133)

The film-makers capture their testimony, and while they may perform a therapeutic role, they typically have other, broader concerns. They are concerned with the bigger picture and the future implications and trajectories of these local instances. As a result they often rely on experts and public figures who are committed to environmental redress but do so from positions of relative distance and privilege. Certainly film-makers and their traumatized filmed subjects share a common urgency to grapple with the truths of their circumstances, which have been hidden, obscured and denied.

A third area of investigation involves a crucial concern—perhaps even a preoccupation—of documentary tradition: truth value. To what extent and in what ways have truth claims been mobilized in the environmental documentary? If these film-makers have been "crafting truth," to take the title of a recent book on documentary, what are the tropes of truth that have been running through the genre and are evident in particular texts? (See Spence and Navarro 2011.) Certainly truth value is not of equal urgency or evenly applicable to all documentaries. Whether Ross McElwee is a "true" counterpart to General William T. Sherman in *Sherman's March* (1986) is a topic that is best confronted playfully. Yet for those documentaries that are concerned with environmental crises, which include localized traumas to discrete ecosystems as well as the looming threat of more global calamities such as global warming, the issue of truth has more immediate and profound relevance. Obviously a systematic pursuit of these three concerns would require book-length treatment rather than a brief chapter. So what follows is at best suggestive and inevitably guilty of gaps, oversights and elisions.

EVOKING THE HISTORY OF ENVIRONMENTAL DOCUMENTARIES

The rapid emergence of the environmental documentary as one of the paramount nonfiction genres has a long pre-history. Scenes of nature's grandeur such as Niagara Falls were among the earliest motion pictures made for projection and shown in theatres. When such films appeared individually on variety programs, they often evoked landscape painting of the nineteenth century; exhibitors, however, also organized these short films into documentary-like, nonfiction programs—then called illustrated lectures—that dealt with man's relationship to nature. E. Burton Holmes had taken numerous photographs during his visit to Yellowstone National Park in August 1896. When he first delivered his evening-length lantern-slide lecture on this subject in New York City (March 1898), Holmes concluded with "a number of motion pictures . . . showing several of the more noted geysers, the ridiculous, bubbling 'paint pots,' and several of the greater falls of the Yellowstone, which called forth warm applause" (Holmes n.p.). The park itself stood in critical relationship to the rapacious exploitation of nature and its natural resources, the foul water and air, the devastated landscapes that resulted from rapid industrialization. These programs probably offered a temporary escape from instances of local ecological trauma rather than a perspective from which to analyze them. They might assure people that pristine nature still existed for potential access and even that some kind of balance between the pristine and industrial development (the sublime and the practical) was in effect.

Holmes's "travel lectures," which survive in book form, can be too quickly dismissed as mere travelogues. In some ways they bear strong resemblance to recent documentaries in which "the documentarist readily becomes an autobiographical essayist who ponders the state of the world on a minimal budget, less interested in simply showing us the world then encouraging us to rethink it" (Romney qtd. in Chanan 12). His *The Yellowstone National Park* mobilized tropes of truth that merit closer analysis. First his photographic images offered a more accurate and detailed view of this wildlife preserve than had been generally available, because magazine and newspapers still largely used lithographic processes for their illustrations. Likewise, his films were more "truthful" than a range of static images in that they could show geysers and other natural spectacles in motion—rather than (at best) suggest such motion. And while photographic images of the park were not unfamiliar, Holmes provided a remarkably elaborated view of the park—a more complete and therefore in some sense more truthful view of the world's first national park—in comparison to other representations. He had also experienced Yellowstone National Park first-hand, which gave his program rhetorical authority. Still difficult to reach by modern means of transportation, the park remained largely hidden to the outside world. Holmes was thus presenting his audience with a world that otherwise remained comparatively unknown and hidden (at one point during his travels through the park, Holmes used a raft to reach a more remote area). Similar kinds of conservationist programs—generally on national parks (often fully integrated slides and films but eventually entirely film)—were frequently offered over the next two decades.

As nonfiction practices developed and changed, subsequent programs became more explicitly part of the documentary tradition. The term "documentary" had become well-established by the mid-1930s. Pare Lorentz's *The Plow That Broke the Plains* (1936) and *The River* (1937), the latter of which won the Best Documentary Award at the 1938 Venice Film Festival, took on environment issues from a new, more critical perspective. They showed images of the devastated American West—the dust bowl and denuded hillsides—that can be said to offer a counter truth, a critical understanding that was absent from the documentary views of national parks. They showed people victimized by these ecological disasters: their farms turned into wind-swept sand dunes, their homes ravaged by floods. Citizens may be shown, but they did not speak (Winston 269–287). The solution—damming the river to control floods and generate electrical power—concludes the latter film. Such themes also resonated through other American documentaries, such as Willard Van Dyke and Ralph Steiner's *The* City (1939) in which the polluted industrial landscape and the unlivable city are featured. The sorry state of the genre in the post–World War II era is best evidenced by the enthusiasm—evident to this day—for Robert Flaherty's *The Louisiana Story* (1948), which celebrated the drilling for oil in the pristine bayous of the Gulf Coast (Barsam 282–285).

Rachel Carson's ground-breaking bestseller *Silent Spring* (1962), which recently celebrated its fiftieth anniversary, not only did much to launch the modern environmental movement, it quickly generated *CBS Reports: The Verdict of the Silent Spring of Rachel Carson* (1963) and renewed documentary's commitment to critical perspectives in this area. The scientific reassurances of experts about the safety of DDT and other chemical poisons used to control pests and diseases were challenged, and new truths about their threat to the ecosystem were offered in their place. It was the depth and analysis of Carson's argument that convinced people that she was discussing what chemical corporations and the like were eager to keep hidden from view. Again she offered a counter truth to their easy-going reassurances.

Although it is important to trace a genealogy of the environmental documentary back into the late nineteenth century and then move forward into the mid twentieth century and beyond, these aforementioned achievements did not constitute a distinct, recognizable genre. Some were associated with the social issue documentary; others with the nature or wildlife documentary, which has flourished over the last eighty years. As Jan-Christopher Horak has explained, wildlife documentaries "were perceived to be an expansion of human vision, a means of entering into a world that was invisible to the human eye, an extension of the physical body of the subject, allowing for the creation of pleasure by bringing animals in their natural habitat closer to humans" (459). Almost all of them depict wildlife existing in an unspoiled state of nature, a world that has an ambiguous relationship to the real world of the spectator.

The establishment of a distinct identity for the environmental documentary involved a shift, reformulation and re-articulation as much as (perhaps more than) the appearance of something entirely new. Such an emergence must be understood in relationship to developments in the environmental movement as well as to documentary and moving-image practices. Earth Day, a day that is intended to inspire awareness and appreciation for the earth's natural environment, was launched in 1970; with it came documentaries such as Lincoln P. Brower's *The Flooding River* (1972), which challenged many of the "command and control" approaches to water management offered by Lorentz's *The River*. It differs stylistically as well. Rather than a narration relying on Lorentz's incantations in combination with Virgil Thomson's music, Brower delivers a dry, scientific explanation on camera in a slightly awkward style that forsakes rhetorical flourishes. It focuses on the Connecticut River Valley, offering a geomorphic-based approach to water control.[1] The trauma experienced by this ecosystem was to be abated by tearing down dams rather than building them. Likewise, in Japan Noriaki Tsuchimoto made the 167-minute *Minamata: The Victims and Their World* (1971), which focuses on the residents of Minamata, many of whom were born deformed or suffered damage to their nervous systems due to the consumption of fish containing abnormal amounts of mercury released into the sea by a fertilizer factory (see Marzani 1972).[2]

Documentaries in the 1980s often focused on the impact and dangers of nuclear radiation and waste, with Judy Irving and Chris Beaver's *Dark Circle* (1982) and Robert Stone's *Radio Bikini* (1988) offering two powerful examples. *Dark Circle* premiered at the 1982 New York Film Festival and won the Grand Prize at Sundance in 1983. *Radio Bikini* also won the Sundance Film Festival Grand Jury Prize and was nominated for an Academy Award. Screened at mainstream festivals, these documentaries were seen as films about an American war machine as well as environmental destruction. They were political documentaries in ways in which Godfrey Reggio's *Koyaanisqatsi: Life out of Balance* (1982) was not. *Koyaanisqatsi*, composed of arresting images of remarkable power, contrasts pristine majestic nature with humanity's violent, devastating abuse of the planet's environment (Figure 2.1). "Koyaanisqatsi," which is a Hopi word meaning "life out of balance," is both the title and the only word that is spoken/chanted in that film. Images of environmental trauma are accompanying by a Philip Glass score.

Environmental film festivals have played a crucial role in the construction of the genre's identity (see de Valck 2007; Stringer 2003; Turan 2003). Although part of a larger film-festival phenomena, their appearance and proliferation coincided with noteworthy events in the environmental movement. Earth Day went international in 1990 and was soon followed by the 1992 United Nations Conference on Environment and Development (UNCED) in Rio de Janeiro—which is often known as the Earth Summit. It is not by chance, then, that the Tokyo Global Environmental Film Festival was launched in 1992, promoting itself as Asia's first international environmental film festival. The Environmental Film Festival in the Nation's Capital (Washington DC), founded in 1993 by Flo Stone, quickly followed, "seeki[ng] to advance public understanding of the environment through the power of film" (DC Environmental Film Festival n.p.). That same year the International Environmental Film Festival (FIMCA) debuted in Barcelona, Spain. The Cine'Eco—International Festival of Environmental Film and Video in Seia, Portugal, has been active since 1995. The Planet in Focus International Environmental Film & Video Festival, based in Toronto, Canada, began in 1999.

Figure 2.1   *Koyaanisqatsi* (1982) presents footage of the 1957 nuclear bomb testing in Nevada, known as Operation Plumbbob.

These festivals provided an important platform for films on the environment—but what kinds of films? When asked to name the five most important environmental documentaries of the 1990s, Stone responded, "It is hard to select just five documentary films from 1993–2000. Here are twelve choices—all very different and in no order" (Stone, conversation n.p.). The following puts her choices in chronological order and offers some additional explication:

- *The Island Sea* (1991), directed by Lucille Carra and Brian Cotnoir, fifty-six minutes, available on DVD. According to the film's promotional material, the documentary "re-creates the lyrical vision of old Japan captured by Donald Richie in his classic travel memoir. Richie, one of the foremost Western authorities on Japanese cinema and culture, juxtaposes the ongoing conflict between traditional and modern values with the serene beauty of the area known as the Inland Sea. The result is a rewarding personal journey for the heart, the mind and the senses."
- *The Spirit of Kuna Yala* (1991), directed by Andrew Young and Susan Todd (streaming online only). As Young describes the sixty-minute film, "A lively portrait of the Kuna Indians of Panama as they unite to protect their homeland, Kuna Yala, and the tradition it inspires. Told entirely in the words of the Kunas, the film contrasts a variety of characters who together tell a story of a culture in flux. At a time when our society is struggling with its relationship to nature, *The Spirit of Kuna Yala* reminds us that the timeless wisdom of indigenous peoples has something vital to offer the Western world" (n.p.).
- *Earth and the American Dream* (1992), directed by Bill Couturié. This ninety-minute HBO documentary won two Emmys and screened at the Sundance Film Festival where it received a Special Jury Award for technical excellence. Critic Marjorie Baumgarten of the *Austin Chronicle* remarked, "The dream is over, and Bill Couturié's provocative documentary about America's blithe destruction of its natural resources sounds the wake-up call. This ambitious film takes on the gargantuan topic of our country's cultural and intellectual history. . . . Beginning with the arrival on these shores of Christopher Columbus, the movie examines the modes of thinking that got us into the ecological dead end we find ourselves in today" (n.p.).
- *Anima Mundi* (1992), a twenty-eight-minute short by Godfrey Reggio. This film was commissioned by the Italian jewellery company Bulgari for the World Wide Fund for Nature, which used the film for its Biological Diversity Program. According to celebrated composer, Philip Glass, who scored the film, "the title *Anima Mundi* reproposes a concept which, throughout the history of mankind from ancient times, conjures up a harmonic principle controlling the laws of life on earth in all its various forms and relationships" (n.p.).

- *Yosemite: The Fate of Heaven* (1990), directed by Jon Else, with narration by Robert Redford, fifty-eight minutes. The documentary is currently available only on VHS. According to Steve Blackburn, "This affectionate history of Yosemite, which was 'discovered' in 1851 as U.S. troops hunted down Native Americans, also raises questions about the future of the park—Americans may be loving it to ruin. Three million visitors per year put a strain on the resources of the park service and physically degrade the region. Highlights include reading from the journal of Lafayette Bunnel, a doctor who accompanied the Mariposa Battalion on its 1851 mission" (n.p.)
- *Vampires, Devilbirds and Spirits: Tales of the Calypso Isles* (1994), directed by Nick Upton for the BBC. A fifty-minute episode for season 12 of "Natural World," this "celebration of Caribbean wildlife . . . takes an entertaining look at the myths and legends of Trinidad and Tobago, and many of the creatures on which they are based" (Vampires n.p.).
- *The Last Frog* (1996), directed by Allison Argo. This winner of several awards at wildlife film festivals such Best of Festival at the Missoula International Wildlife Film Festival has become virtually invisible. *The Last Frog* is 25 minutes long, and was never released on DVD (or VHS) by Nat Geo due to its length. As Argo describes the film, "*The Last Frog* is a love letter to the frog—and a strange and unnerving murder mystery. Starring our amphibious friends, the drama is supported by a cast of scientific experts and devotees frantically trying to unravel the disturbing puzzle behind the curious decline of frogs" (n.p.).
- *Microcosmos* (1996), directed by Claude Nuridsany and Marie Perennou, eighty minutes, available on DVD with a companion book by the film-makers: *Microcosmos: The Invisible World of Insects* (1997).
- *The Saltmen of Tibet* (1997), directed by Ulrike Koch, 108 minutes, available on DVD. It is worth noting that this film won best film at a number of modest-sized but respected film festivals and had a small theatrical release. Stephen Holden of the *New York Times* found it to be a "profoundly absorbing study of a small band of nomads trekking across the Tibetan high country with a caravan of 160 yaks, the camera draws back to observe from afar an annual pilgrimage that has been taking place for 2,000 years" (Holden n.p.).
- *Vision Man: An Eskimo Hunter* (1998), directed by William Long and Lars Aby. This 52-minute, Swedish documentary centres on Utuniarsuak Avike, "an 87-year-old native hunter in northwest Greenland who has spent much of his life literally walking on thin ice. Now confined to a small apartment, Avike tells of the life that lies behind him: fishing, dogsledding, coexisting with wolves and walruses and the ritual religious hunting of the polar bear. Director William Long punctuates Avikes's recollections with reenacted scenes that take in the overwhelming and humbling beauty of the Arctic (both above and

under water) and its wildlife. Not only does this vision man share his well-spent life but he also sheds light on a culture steadily eroding in this changing world" ("Vision Man" n.p.). It was shown in the US and recently became available for educational institutions through Alexander Street Press.

- *A Place in the Land* (1998), a thirty-two-minute Oscar-nominated short directed by Charles Guggenheim. This film "examines the history of conservation stewardship in America as it is reflected in this property and through the work of George Perkins Marsh, Frederick Billings, and Laurance S. Rockefeller, successive residents of the estate [i.e. the Marsh-Billings-Rockefeller National Historical Park]" ("A Place in the Land" n.p.).
- *Charcoal People* (2000), directed by Nigel Nobel. A sixty-five-minute "documentary about the rural population who earn their living as coal miners, thus helping to keep metallurgic activity going and contributing to the forest devastation in South America" ("Charcoal People" n.p.).

Stone added that, "It is important to think about TV documentaries during those years including the great increase in wildlife programming led by the BBC and the influence of IMAX" (Stone n.p.).[3]

Although these documentaries often hint at the more radical and politically engaged environmental documentaries that were to come, perhaps only *Earth and the American Dream* directly addresses issues on a scale and in a manner that have become relatively familiar. Although released the year of the Environmental Summit, Couturié's effort was not always appreciated. Ken Tucker of *Entertainment Weekly* dismissed the film-maker as a "tree-hugger" and found the documentary to be "sincere but boring, a warning that comes off as condescending" (Tucker n.p.). It was, however, the first documentary to win an Environmental Media Award. Most of the other documentaries seem quite traditional by comparison. At least two (*Yosemite: The Fate of Heaven* and *A Place in the Land*) are about national parks in a manner that goes back to Burton Holmes's late-nineteenth-century programs on Yellowstone. They take a Rooseveltian (Theodore), conservationist approach. Several others are ethnographic in emphasis, focusing on peoples whose traditional lifestyles are under threat if not rapidly disappearing. *Vision Man: An Eskimo Hunter* cannot but recall Flaherty's *Nanook of the North* (1922) and *The Saltmen of Tibet* brings to mind Miriam C. Cooper and Ernest B. Schoedsack's *Grass: A Nation's Battle for Life* (1925), whereas *The Spirit of Kuna Yala* seems in the tradition of Robert Gardner's *Dead Birds* (1965). Others, *An Inland Sea* in particular, participate in the long-standing travel genre. Many are wildlife documentaries—a genre that boasted quite a few film festivals in this period, some of which had been running since the 1970s.[4] Simply put, Stone's list suggests that environmental documentaries of the 1990s were scattered across a number of established genres that could be mobilized to reveal dangers to peoples, habitats and fauna—pointing towards the larger environmental challenges facing the world. In often small but precise ways, they revealed developments that were largely hidden and sought to foster an informed citizenry in the tradition of John Grierson (1966).

Although the environment was a pressing issue in the 1990s, the most dynamic and important documentary genre in the US during this period proved to be the courtroom documentary with its focus on legal film truth—beginning with Errol Morris's *The Thin Blue Line* (1988) and Chris Choy and Rene Tajima's *Who Killed Vincent Chin?* (1988; see Johnson 1996; Nichols 1996). Their methods and approaches provided a general framework for more than twenty major documentaries, including Joe Berlinger and Bruce Sinofsky *Brother's Keeper* (1992) and *Paradise Lost: The Child Murders at Robin Hood Hills* (1996), Nick Broomfield's *Aileen Wuornos: The Selling of a Serial Killer* (1993), Choy's *Shot Heard 'Round the World* (1997), and Morris's own *Mr Death: The Rise and Fall of Fred A. Leuchter, Jr.* (1999). It continued into the new century with Jean-Xavier de Lestrade's *Murder on Sunday Morning* (2001) and *The Staircase* (2004), as well as Andrew Jarecki's *Capturing the Friedmans* (2003).

In the first half-decade of the new millennium feature-length documentaries that dealt with environmental issues focused on an array of subjects and used a variety of styles. They included Agnes Varda's *The Gleaners and I* (2000), Judith Helfand and Daniel B. Gold's *Blue Vinyl* (2001), the last of Godfrey Reggio's poetic Qatsi trilogy—*Naqoyqatsi* (2002)—Hubert Sauper's *Darwin's Nightmare* (2004), Morgan Spurlock's *Supersize Me* (2004), Martin Marecek's *The Source (Zdroj,* 2005) and Michael Glawogger's *Workingman's Death* (2005).[5] In contrast to many of the 1990s documentaries listed by Flo Stone these were feature length, received extensive critical attention and generally enjoyed theatrical distribution (even if limited and not always in the US). More than those documentaries of the previous decade, they explicitly engaged large-scale aspects of the environmental crisis. Nevertheless, their diversity of subject matter and approach—of semantics and syntax—poses the question: Did they then constitute a coherent genre? (Altman 24). Moreover, attention and perhaps even the genre's maturation were put on hold by a host of documentaries engaging the many urgent issues that followed the World Trade Center attacks on September 11, 2001: the wars in Iraq and Afghanistan, the "war on terror" and the conduct of the Bush administration as exemplified by Robert Greenwald's *Uncovered: The Truth about the Iraq War* (2003), Michael Moore's *Fahrenheit 9/11* (2004) and many others.

The full constitution of the environmental documentary as a genre coincided with a second wave of new environmental film festivals around the world. These include EcoCinema, Jerusalem's International Environmental Film Festival; the Green Film Festival in Seoul (South Korea); and the San Francisco Ocean Film Festival: all were established in 2004. The New Zealand–based Real Earth Film Festival was founded in 2005. In 2007 the

Eugene P. Odum School of Ecology at the University of Georgia inaugu-
rated the EcoFocus Film Festival and the Princeton Public Library began the
Princeton Environmental Film Festival. The Environmental Film Festival at
Yale, affiliated with its School of Forestry and Environmental Studies, began
in 2009—as did the Environmental Film Festival of Accra, Ghana. This proliferation certainly reflected the increasing centrality and dynamism of the
environmental documentary.

AN INCONVENIENT TRUTH (2006)

Davis Guggenheim's *An Inconvenient Truth* (2006) was a key film in the
genre's solidification and rise to prominence. Flo Stone of the Environmental Film Festival in the Nation's Capital has characterized it as "a decisive
moment" for environmental documentary: After its release, "no one asked
us anymore what we were trying to do and if anyone would want to come"
(Stone, conversation, n.p.). The film bears a striking relationship to Reggio's
Koyaanisqatsi. One might say that "life out of balance" is the "inconvenient
truth" of Guggenheim's documentary, although it is one is done as a visual
poesis, whereas the other is a discourse-heavy popular science presentation.
Like *Koyaanisqatsi, An Inconvenient Truth* opens with scenes of pristine
nature (in this instance, a river that runs along Al Gore's farm) and soon
shifts to a litany of images depicting ecological devastation and eco-trauma
(from bodies of people who died in Hurricane Katrina through melted glaciers). Scientists and politicians appear in the film but generally lack a voice,
their insights or failings being characterized by Al Gore's narration. If we see
people caught up in environmental calamities, they are never given a chance
to speak. People's trauma, in the past and in the future, is left to our imagination. In this respect, the sequences of images of New Orleans in the wake of
Hurricane Katrina stand in marked contrast to Spike Lee's *When the Levees
Broke* (also 2006), in which residents speak about the physical and psychological trauma that they, their family and fellow residents have endured [7].

   An Inconvenient Truth shared one significant characteristic with many of
the Iraq-war documentaries of the same period: not only politically partisan
(stock footage clips show Republicans to be climate change deniers), it was
designed to intervene in the electoral process. Its counterpart in this regard
was Robert Greenwald's *Iraq for Sale: The War Profiteers* (2006). After
premiering at Sundance in January, *An Inconvenient Truth* went on to gross
over $24 million during its theatrical run from May 24 to November 2,
2006—a period that coincided rather perfectly with the off-year campaign
season. It interweaves a biography of former Vice President Al Gore, the
Democratic presidential candidate who had lost to George Bush in 2004,
and a re-presentation of his frequently given PowerPoint lecture on global
warming. At several points the film takes aim at his nemesis—President
George W. Bush. Even as *An Inconvenient Truth* helped Democrats wins

some seats in the US House of Representatives and Senate, it further polarized the issue of global warming in an already highly charged partisan environment. Certainly the documentary won few converts among Republicans: If anything, the associations of this environmental issue with Al Gore and the Democrats inclined many of them to dismiss global warming.[6]

At the same time, *An Inconvenient Truth* was able to transcend its particular political moment, in part because of the importance of its topic and also through its creative mobilization of certain tropes, certain methods of presenting truth that have been part of the documentary tradition. That the issue of truth was raised by the title is significant in this light. Truth, for Gore, is a scientific truth, based on evidence and tested relations between cause and effect. In particular, rising levels of CO^2 are shown to be major contributors to global warming. People—citizens and politicians—consciously and unconsciously want to avoid these scientific realities. As Gore asserts, "There are good people who are in politics in both parties who hold this at arms length because if they acknowledge it and recognize it, then the moral imperative to make big changes is inescapable." In one section of the film, a young Al Gore is seen questioning a NASA scientist who admits that the final paragraph of what he delivered was not written by him and did not reflect his scientific assessment. This is followed by a section in which statements by a scientist are taken out of a policy document because their conclusions did not conform to the Bush administration's position.

Truth is normally juxtaposed to lies. Moreover, in an earlier cycle of prominent documentaries, a film's counter truth confronts a state truth, which is shown to be a lie. In *The Thin Blue Line*, Randall Adams is guilty of murder—this is a state truth for which he was to pay with his life. The film shows that this truth is a lie and goes so far as to identify the actual killer, who had gone on to murder again. With *An Inconvenient Truth* there is not so much a state truth as a state doubt. The state, which is to say the then current Republican administration, asserts that the evidence for determining the causes of global warming is inconclusive. These administrations offer other possible explanations (e.g., a cycle of climate variation). In the meantime they suppress and rewrite the statements of scientists working for government agencies and have industry spokesmen run the Environmental Protection Agency (EPA). The state's reliance on ambiguity and uncertainty is an effective strategy of obscuration. Complexity and ambiguity is normally the domain of liberals. In short, the lack of a clear truth—of scientific certainty—becomes the state endorsed truth. The goal of *An Inconvenient Truth* has been to confront these doubts (the state truth) and show that they do not exist. It is thus important to assert that no peer-reviewed articles in scientific journals deny the reality of global warming. In short, Gore and Guggenheim strive to show that the state's truth of doubt is a lie.

The figurations of truth are not unfamiliar. In the scene discussed earlier, the assessments of a government scientist were removed by an administration lackey, who lacked any scientific qualification. Not the substance but the act of deletion is revealed, given the light of day. In another scene, Gore works with the military to release state secrets that document the rapid shrinkage of the polar ice caps since 1970. In addition—and this is one of the moments in the film when rhetoric and aesthetic pleasure seem to coincide—the bringing to light of the state secrets is combined with the surfacing of the submarine as it breaks through the ice pack. It is a complex metaphor in that the breaking through of the sub is itself a kind of proof that something is wrong. The submarine's coming into the light contains a certain irony. The cumulative force of these individual revelations is to offer a more complete picture—something closer to the full story, which makes audiences for this film feel that their previous understanding of the issues was incomplete. That is, our former understanding about the state of the environment (what might be considered our personal truth—what is true for us) is felt to have been inadequate and a new, more complete truth emerges. Interestingly, Robin Murray and Joseph Heumann see environmental nostalgia—the eco-memory it evokes—as the key to the film's rhetorical success. This strategy, however, depends on making the audience aware of current trends in climate change—and the difference between then (circa 1970) and now as well as between now and the future (Murray and Heumann n.p.).

For the purveyors of doubt the strategy to contest this documentary is simple—to look for weaknesses (exaggerations, instances of overreaching), mistakes or lies in the facts and arguments that Gore and Guggenheim present in the film. A court case in England provided a significant opening when a judge found nine "errors" in the documentary. These so-called errors are not to say that Gore is wrong but that there is a lack of consensus, and other credible viewpoints exist. A range of opinions exists about what is happening, and Gore in some cases chose some questionable examples. To cite one "error," scientists suggest there are various possible contributing factors to the melting of the glacier on Mt. Kilimanjaro of which global warming is only one. Gore merely chose a bad example because there is much less dispute about global warming's impact on other glaciers and ice masses.

Detractors focused on Gore contention that the earth's water level was in danger of rising twenty feet in the near future, which the judge declared to be unduly alarmist (Figure 2.2). Doubters used this as an opening. H. Sterling Burnett of the National Center for Policy Analysis, which has been heavily funded by ExxonMobil, argues,

> What Gore doesn't say about the threat to the ice sheets is as important as what he does say, however. Ice and snow is accumulating in the interior of Greenland and Antarctica, but decreasing around the edges. A 2005 study in the *Journal of Glaciology* by a NASA scientist concludes that there is a net loss of ice that will result in higher sea levels. But the loss is occurring slowly: 0.05 millimeters on average per year. At

that rate, it will take a millennium for the oceans to rise 5 centimeters (roughly 2 inches) and 20,000 years to rise a full meter. More recent research indicates that the pace of melting has increased. But even under the worst case it would take at least several centuries—1,800 years by one calculation—for the scenario painted in the movie to play out, giving humans a considerable amount of time to adapt.

(n.p.)

Many of the methods used to challenge *An Inconvenient Truth* had been previously developed and applied to Michael Moore's *Fahrenheit 9/11*. Al Gore was now in the Michael Moore position. The goal was to establish Al Gore as an unreliable narrator and re-establish a sense of doubt and uncertainty, which would enable the US to continue dodging this issue. As Nichols notes, to the extent to which a speaker's "good name" has been undermined or destroyed, a documentary lacks credibility and truth value (*Representing* 135).

Gore's position is perhaps close to that of James Hansen, who heads NASA's Goddard Institute for Space Studies. Hansen argues that there is tremendous pressure on scientists from the funding agency to be conservative in their estimates of future climate change. Similar kinds of pressure and selective mining of data provided the cover for Bush's invasion of Iraq. Nonetheless, Hansen writes, "I find it almost inconceivable that 'business as usual' climate change will not result in a rise in sea level measured in meters within a century" (n.p.). Catherine Brahic, an environmental reporter, sought a balanced assessment—a middle ground. For instance, disease is an

Figure 2.2 Al Gore presents this digitally fabricated image of Manhattan suffused with water in *An Inconvenient Truth* (Davis Guggenheim 2006).

important factor in the bleaching of coral—global warming is only an indirect factor by contributing to a changing environment that fosters disease. On the other hand, if a significant change in temperature in the oceans does occur, it will bleach the coral reefs. Brahic thus concludes that

> strictly speaking, Gore oversimplified certain points, made a few factual errors and, at times, chose the wrong poster child (Mount Kilimanjaro should have been replaced by any number of Alaskan or Andean glaciers, for instance). It's unfortunate, but it remains the most comprehensive popular documentary on climate change science I have seen.
>
> (n.p.)

The opposition has continued. Julia A. Seymour, an assistant editor and Analyst for the Business & Media Institute of the Media Research Center "analyzed broadcast news coverage of Gore about climate change and mentions of *An Inconvenient Truth*" over a five-year period through the end of April 2011. Among her conclusions were that "the networks shouldn't take his interpretation of global warming science as truth. Rather, they should be skeptical because of [Gore's] very real political agenda" (Seymour n.p.). Any discussion of climate change needed to be "balanced" to include the opinions of climate change sceptics.

An Inconvenient Truth was not the only environmental film at the 2006 Sundance Film Festival: another was Chris Paine's *Who Killed the Electric Car?* (2006). However, Iraq War documentaries were still getting the bulk of attention, including Patricia Foulkrod's *The Ground Truth* and James Longley's *Iraq in Fragments*, which won three awards at the festival. *Who Killed the Electric Car?* reveals the history of the EV1 (Electrical Vehicle) cars in California in the late 1990s and early 2000s—a history that automobile and oil companies have done their best to conceal, even to the point of repossessing and destroying all the EV1s on the road. Moreover, it shows that the California Air Resources Board head Alan Lloyd became director of the California Fuel Cell Partnership a few months before voting down the regulations that had supported the commercial introduction of Electric Vehicles. Hydrogen fuel cells were promoted as the new alternative to gasoline-powered automobiles, generating over a billion dollars in funding and subsidies. If the principle villains in this story are the auto industry and oil companies, the documentary shows a number of scenes that focus on George W. Bush and his administration's contributions to the successful execution of the electric car. In this respect it shares an anti-Bush/anti-Republican agenda with *An Inconvenient Truth* and *Iraq for Sale*. Its theatrical run—opening June 28, 2006, and concluding November 15, 2006—can also be seen as geared toward impacting the off-year elections. Given its box office of over $1.6 million, the film had significant visibility.

The Environmental Media Association, like Flo Stone, sees 2006 as a key turning point in which environmental media contributed to a "new norm," stating,

> In 2006, environmentalism achieved a tipping point. Between the impact of Al Gore's documentary, *An Inconvenient Truth*, and a series of natural disasters, environmentalism went from being the work of activists to an every day concern for regular people. Now more than ever, people are talking—*and doing something*—about the environment. Conserving energy, buying locally and simply thinking more about consumerism overall are some ways people are going green.
>
> ("EMA Awards" n.p.)

Besides the two documentary features already mentioned, HBO offered *Too Hot Not to Handle* (2006), also about climate change. Laurie David, who was one of the producers on both *Too Hot Not to Handle* and *An Inconvenient Truth*, also published *Stop Global Warming: The Solution Is You!* described by one environmental website as "a handy pocket guide to curbing climate change" (Dunn n.p.). In the same year, the National Film Board of Canada offered Jennifer Baichwal's *Manufactured Landscapes* (2006), a documentary portrait of photographer Edward Burtynsky, which "shares Burtynsky's astonishment and concern over the scale, tempo and irreversibility of postmodern humanity's global frenzy of production and consumption" (Baker n.p.). It is a powerful and discerning look at an artist who confronts the world in the midst of a growing environmental crisis (Cammaer 121–130).

ENVIRONMENTAL DOCUMENTARIES AS BUSH LEAVES OFFICE

By 2008, a wave of documentaries was engaging many facets of the complex, multifaceted environmental crisis. Irena Salina's *Flow: For Love of Water* premiered at the 2008 Sundance Film Festival and had its theatrical run as a ninety-three-minute feature from March to December 2008, during the electoral campaign season. Its box office was a modest $142,569, and was recut and released on DVD as the eighty-four-minute *Flow: How Did a Handful of Corporations Steal Our Water?* (released December 11, 2008). It was the principal environmental documentary to be released theatrically during the presidential election year—when commercially successful documentaries such as Larry Charles' *Religulous* with Bill Maher, Scorsese's *Shine a Light* with the Rolling Stones, and Stephen Walker and Sally George's *Young@Heart* offered counterpoints to media-saturated, politically charged discourses of election year politics.

Gestures toward partisan party politics, which had turned *An Inconvenient Truth* into a lightning rod, were increasingly avoided by documentary film-makers and distributors tackling hot-button issues such as the environmental crisis. It is worth noting that Michael Moore released *Sicko* and *Capitalism: A Love Story* in 2007 and 2009, respectively—non-election years. Likewise the other prominent environmental documentary at the 2008 Sundance Film Festival was Josh Tickell's *Field of Fuel*, which won the Audience Award for Documentary. Tickell recut the film and released it theatrical in September 2009 and on DVD in 2010. The normal cycle of festivals and theatrical release may have been one factor, but those environmental documentaries screened at festivals during the closing months of the Bush administration and beyond were not released commercially until Obama was in office—when a new administration would be more sympathetic to the makers' environmental aspirations. Of course, it was not clear who would be president when the films were being made, but these films were not screened in the midst of a political campaign.

This next wave of documentaries, which tend to see corporations more than the government as the likely villains, often focused on specific instances of environmental destruction and the traumatic impact, both physical and psychological, it is having on people. Joe Berlinger's *Crude* investigates an area of once-pristine rainforest in Ecuador that was decimated by oil extraction—an area that has been called an "Amazon Chernobyl" (Mcavoy n.p.). It also engages many of those suffering directly from this disaster as well as organizers and lawyers seeking justice on their behalf. This includes an indigenous woman, who begins the film by singing a song in a high, sweet voice:

> We lived upon the river of rich clear waters.
> With the arrival of the company and their contamination
> my brothers are now dead.
> I am the only survivor of my family.
> The message of my song is to tell the world
> so that the world can know what has been done.
> I worry about the future.
> What will happen to the children?

Members of the Secoya people soon speak about the rainforest that was undamaged before Texaco arrived. Berlinger focuses on a class action suit against Chevron/Texaco, demanding that the company accept moral and financial responsibility for its contribution to this oil-related disaster—which Chevron dismisses as a story made up by con men wanting to enrich themselves. This is the corporate truth that Chevron defends in the court and seeks to make a state truth. As the courtroom battle unfolds, *Crude* gradually reveals the real cost of pumping oil. In this respect, Berlinger continues his focus on legal truths from his series of Paradise Lost documentaries, which focused on the scenarios of murder with the West Memphis Three.

Two environmental documentaries from this period that let traumatized individuals bear witness while mobilizing tropes of truth are *Food, Inc.* and *The Cove*. Testimony as a ritual of healing has been widely noted. As Judith Herman has remarked, "Testimony has both a private dimension, which is confessional and spiritual, and a public aspect, which is political and judicial" (181). Robert Kenner's *Food, Inc.* takes aim at "factory farming" and sees large, impersonal corporations—driven by the goal of ever increasing profits—as the principal culprit. As one voice (who remains unidentified) remarks, "The companies don't want farmers talking. They don't want this story told." One section looks at chicken farming, which increasingly is conducted in large darkened barns. A farmer is eager to show the film-makers how he raises his chickens but Tyson, the corporation for which he works, convinces him not to do so. Tyson wants us—like the chickens—to be in the dark. Many farmers remain silent, but finally one Carole Morison has reached a point where she feels she cannot continue. What she is doing "isn't right" and so she is eager to talk and let the film-makers record what goes on: Despite the heavy use of antibiotics, some chickens die in her barns—and she herself has become allergic to antibiotics. Likewise, the film-makers use night vision cameras to videotape the "harvesting" of chickens, which emphasizes the fact that they need to act surreptitiously to show what was actually happening. They are showing what is behind the corporate image—an unmasking.

Perdue and Tyson try to stonewall the film-makers. Elder Roth, the founder of Beef Products, Inc., took a different approach and let them into his high-tech world of mammoth machinery and centralized control where he shows how his sanitary strategies—the heavy use of ammonia—battles the growing dangers of E. coli in food. Despite Roth's cooperation with the film-makers, the underlying trope in this scene remains strikingly similarities to those involving Perdue. We are shown what the walls of the building obscure—a horrific world of slaughter where food (meat) is produced with indifference to animals and consumers alike. Likewise, in considering the people employed at Smithfield, the film-makers give hidden cameras to the workers so they can capture what takes place inside the processing plant. *Food, Inc.* tries to show the ways in which the individual components are interlinked. If Tyson tries to keep viewers in the dark like its chickens, Smithfield treats its employees the same way it treats its hogs. While *An Inconvenient Truth* seeks to offer a comprehensive, integrated understanding of global warming, *Food, Inc.* does much the same for our food system. However, its methods differ substantially. There is no single voice of reason like Al Gore, although the film-makers have their own narration voice; rather it works primarily by integrating the voices of journalists such as Eric Schlosser and Michael Pollan, who have investigated food-related issues, and the people who are being chewed up by the food factory system. Although corporations are able to silence many of those trapped in the system they have done much to create, they are not able to silence everyone.

The experts offer the film-makers authoritative voices that provide facts, contextual understanding and rational argument. The traumatized who finally speak provide the film's emotional rhetoric. But for some who have been traumatized, such as Barbara Kowalcyk, the need to talk and act has turned them into experts. Kowalcyk's six-year-old son died from E. coli infection that came eating contaminated hamburgers. Not only the farmers and those processing food on assembly lines but consumers very much like ourselves are all participating in a system of eco-trauma. The film pursues any number of strategies for revealing the truth of what these corporations don't want us to know about this system.

Louie Psihoyos's *The Cove* focuses on the annual killing of roughly 23,000 dolphins and porpoises in Japan each year—many in a cove in Taiji (a small number of these mammals are actually captured and sent to theme parks). Using some of the same methods for crafting truth, the film argues that exposure—tearing away the veil of secrecy—is the crucial first step towards ending this slaughter. The film-makers then go on a high-tech adventure to capture compelling audio-visual materials of the brutal killing—assembling a team of activists, film-makers and divers whom they compare to the characters in the Hollywood caper film *Oceans Eleven*. Under the cover of darkness they slip into the forbidden area, their actions recorded by night cameras. Members of the team serve as the film's protagonists as they evade the police. Finally we get to see what they captured with their hidden cameras—the casual conversations of the fishermen as well as the dolphin killings. They have outwitted the fishermen just as the fishermen have outwitted the dolphins in order to drive them into their nets. All this would be problematic in various ways—at times it comes a little close to an Osa and Martin Johnson filming expedition in Africa or TV docs with their on-camera animal wranglers. However, Psihoyos provides other components that complicate and arguably rescue the film.

The rhetorical success of *The Cove* depends on the people as well as the animals who are traumatized. One key member of the film crew is dolphin trainer Ric O'Barry, who worked on the TV show *Flipper*. Through his close work with the dolphins O'Barry gradually realized that he was participating in a system that was cruel and corrupt. By training dolphins, he came to recognize their extraordinary intelligence that makes them cognizant beings. He also realized he was participating in their exploitation and death. O'Barry reacted by becoming an outspoken "dolphin defender." The weight and recognition of what he did, compels O'Barry to speak out—and to act.[8] The film concludes first with O'Barry defiantly entering a meeting of the International Whaling Commission and showing scenes of the dolphin slaughter on a large portable video screen, disrupting the commission members self-congratulatory pronouncements of "shortening the time to death." Finally, in a gesture of self-abjection, he stands in the streets of Tokyo with the same video screen showing the images to passing pedestrians.

It is likewise significant that the fishermen and their advocates have been deceiving the Japanese public not only by keeping what they are doing a secret but by selling highly toxic dolphin meat as whale meat from the pure waters off South America. DNA testing—used to exonerate unjustly convicted criminals in the US court system—is mobilized to expose this deceit. Thus the Japanese have been the real victims of this practice given the elevated levels of mercury in their bodies—a form of toxicity that resonates with the Noriaki Tsuchimoto's documentaries of Minamata. In one telling scene, the government spokesman who tries to justify the practice allows the film-makers to clip a piece of his hair. This enables them to diagnose him as suffering from mercury poisoning. The fishermen are not only killing dolphins and harming Japanese gourmands; as consumers of dolphin meat, they are killing themselves and their children.

The environmental documentary continued to flourish in 2010 as two examples received nominations for an Academy Award as Best Feature-Length Documentary: Lucy Walker's *Waste Land* and Josh Fox's *Gasland*. *Gasland* earned the ire of the oil and gas industry, which actively campaigned to prevent the film from winning the Oscar. Fox's ground-breaking investigation of hydraulic fracturing (known as "fracking") and its impact on the environment effectively employs an array of strategies for presenting truths. The film opens with representatives from leading natural gas corporations making statements before the Subcommittee on Energy and Minerals:

> Studies and surveys by GWPC, EPA and IOGCC over the last eleven years have found no real credible threat to underground drinking water from hydraulic fracturing.
>
> Recently, however, there's been concern raised about the methods to tap these valuable resources. Technologies such as the practice of hydraulic fracturing have been characterized as environmentally risky and inadequately regulated. Press reports and websites alleging that six states have documented over one thousand incidents of ground water contamination resulting from the practice of hydraulic fracturing. Such reports are not accurate.
>
> It is my firmly held view as well as that of IOGCC that the subject of hydraulic fracture is adequately regulated by the states and needs no further study.

The remainder of the film effectively offers testimony and visual evidence that contradict these pronouncements in order to present a counter truth. Some of the most compelling is from ordinary Americans who suddenly found that their way of life had been fundamentally disrupted by the effects of fracking. Fox begins his investigation by visiting various residents of Dimock, the ground zero for hydraulic fracturing in Pennsylvania with forty wells. Their water has gone bad: They and their animals were suffering unexplained illnesses and other side effects. They are disoriented, angry and mostly eager to talk. Later Fox talks to Mike Markham and Marsha Mendenhall of Fort Lupton, Colorado, who set fire to the water coming out of their kitchen faucet. Others can perform the same trick as well. According to one informant, the health of many of these people had been ruined and they can no longer function.

An array of experts such as Weston Wilson of the EPA is burdened by a sense of frustration and guilt that their findings have been suppressed. They are haunted by secrecy and the overweening power of the oil and gas industry. Dr. Theo Colborn, a renowned environmental health analyst who has done some serious investigation, details health effects that include irreversible brain damage. Experts in Fox's documentary such as air quality specialist Dr. Al Armrendariz and environmental scientist Wilma Subra seem deeply disturbed by their findings. They are little different from those ordinary victims—harassed, disempowered, anxious. Fox's mobile camera and disjunctive editing techniques powerfully evoke the disorientation of his traumatized informants, showing his remarkable insights into their dilemmas and despair. In this respect he has found an expressive form that is adequate to his subject matter. It not only reflects the psychological state of the people with whom he speaks but his own deepening sense of nightmare as he traverse the US only to discover that he is actually descending into a kind of hell.

The corporate truth presented to Congress by industry representatives at the beginning of the *Gasland* must either be accepted and made into an ongoing state truth or challenged so that oversight, study and appropriate restrictions can be imposed. Underlying Fox's citizen's investigation was the stonewalling of corporations that refused to be interviewed or did not respond to phone calls—hoping perhaps that the documentary would not appear or have little visibility. When *Gasland* gained visibility and its counter truth gained credibility, the natural gas industry responded by claiming that Fox's documentary was alarmist and misrepresented the impact of drilling in several specific instances—proving its overall undertaking to be deceptive. In one instance, John Hanger, the secretary of the Pennsylvania Department of Environmental Protection, is quoted saying that the film was "fundamentally dishonest" and "a deliberately false presentation for dramatic effect" ("*Gasland* Debunked" n.p.). But Fox's documentary does not claim to have thoroughly investigated the cause for the many incidents it presents. Rather it refutes the industry's claims that hydraulic fracturing is benevolent and requires no further serious study and oversight.

Although the industry seeks to demonstrate that several key instances in Fox's documentary are inaccurate and part of an overall pattern of deception, their assertions prove questionable or at least not beyond dispute. America's Natural Gas Alliance asserts,

> In the film's signature moment Mike Markham, a landowner, ignites his tap water. The film leaves the viewer with the false impression that the flaming tap water is a result of natural gas drilling. However, according

to the Colorado Oil and Gas Conservation Commission, which tested Markham's water in 2008, there were "no indications of oil & gas related impacts to water well." Instead the investigation found that the methane was "biogenic" in nature, meaning it was naturally occurring and that his water well was drilled into a natural gas pocket.

This is one of several examples where the film veers from the facts.

("*Gasland* Debunked" n.p.)

The commission concludes that Markham's water problems are unrelated to fracking:

> Dr. Anthony Ingraffea, D.C. Baum Professor of Engineering at Cornell University, whose research has involved fracture mechanics for more than 30 years, has said that drilling and hydraulic fracturing can liberate biogenic natural gas into a fresh water aquifer. That is, just because gas is biogenic does not necessarily indicate that it reached a well by natural means.
>
> (Fox n.p.)

America's Natural Gas Alliance also produced a short video entitled "The Truth about *Gasland*." The piece is filled with shots of happy children and reassuring platitudes.

In confronting the growing environmental crisis, documentary film-makers have been taking on not just individual corporations but whole industries—even groups of industries. They reveal their devastating impact on the environment and often give voice to people suffering from the worst effects of this environmental destruction. These are people who seemingly have little or nothing left to lose—who now speak truth to power because other efforts to ameliorate their situation have failed. By bearing witness they offer both a psycho-analytic and political truth by revealing what has been hidden, suppressed or dismissed. In these films—*An Inconvenient Truth*, *Food, Inc.* and *Gasland* are just prominent examples—the rhetoric of truth has played and will continue to play a crucial role.

POSTSCRIPT

Numerous environmental documentaries have been produced and shown at festivals in the early 2010s, including Josh Fox follow-up *Gasland 2* (2013). An inventory of these films can be found through the websites of various environmental film festivals that continue to flourish. Only one such feature-length documentary, *If a Tree Falls: A Story of the Earth Liberation Front*, received an Oscar nomination in 2011. No such documentaries were nominated in either 2012 or 2013. The genre remains a productive and important one but only one of many presently competing for our attention.

Does this mean that the genre has reached a certain maturity and awaits new approaches or new crisis to regain these forms of prestigious but fickle visibility? Or has the broad citizenry once again come to accept the growing environmental crisis with some combination of denial, small symbolic gestures and/or despairing resignation as they seek distractions in other topics such as Miley Cyrus and twerking?[9]

NOTES

1. The assumption was that mankind can dominate rivers for human good. Thanks to Jim MacBroom, Yale School of Forestry and Environmental Sciences, for his reflections on these films, which I have incorporated here. See also United States Environmental Protection Agency, "Hydrologic/Geomorphic Assessments," http://water.epa.gov/polwaste/nps/watershed/hydrologic_geomorphic.cfm

2. *Minamata* is now available on DVD from Zakka Films. See http://zakkafilms.com/

3. Stone also added that, "This was also a time when very special animated films about the environment were being made: *The Mighty River* as well as *The Man Who Planted Trees* by Frederic Back and *Turtle World* by Nick Hilligoss."

4. The International Wildlife Film Festival, established in 1977, claims to be "the first juried wildlife film festival in the world" (http://wildlifefilms.org/about/). While wildlife topics and environmental issues can overlap, they obviously differ.

5. If wildlife films are added to the list, one might want to include *March of the Penguins* (2005) and Werner Herzog's *Grizzly Man* (2005).

6. My analysis suggests that *An Inconvenient Truth* was more divisive than Mark Minster argues in his otherwise productive "The Rhetoric of Ascent in *An Inconvenient Truth* and *Everything's Cool*," in Paula Willoquet-Maricondi, *Framing the World: Explorations in Ecocriticism and Film*. Charlottesville: University of Virginia Press, 2010. 25–42.

7. Spike Lee's *When the Levees Broke* (2006) certainly has an environmental component but the documentary engages other issues, of which the most important is probably the long-standing indifference to the lives of African Americans and the poor.

8. This is also the case for another crew member, Hardy Jones. Hardy has a more modest role in this film, which is not entirely surprising given that he is featured in *The Dolphin Defender* (2005), a television documentary done for "Nature" through WNET, which possesses many of the tropes and some of the same personnel of *The Cove*.

9. Miley Cyrus was the most popular search query for 2013 according to Yahoo ("Stars").

WORKS CITED

"2008 Toronto Film Festival." Canada.com, 28 August 2008. www.canada.com/topics/news/national/story.html?id=9e40c848-c4e6-4852-a99c-7785b541392a. Web. November 2013.

Altman, Rick. *Film/Genre*. London: BFI, 1999. Print.

American Natural Gas Alliance. "The Truth about *Gasland*." 21 January 2012. www.anga.us/media/content/F7D1441A-09A5-D06A-9EC93BBE46772E12/files/anga%20gasland%20truth%20formatted.pdf. Web. November 2013.

Argo, Alison. "The Last Frog." January 2011. www.argofilms.com/film/last-frog. Web. July 2014.

Baker, Kenneth. "Manufactured Landscapes." *San Francisco Chronicle*, 20 July 2007. www.sfgate.com/cgi-bin/article.cgi?f=/c/a/2007/07/20/DDGHQR324E1.DTL#flick. Web. November 2013.

Barsam, Richard. *Non-fiction Film: A Critical History*. Revised and expanded edition. Bloomington: Indiana University Press, 1992. Print.

Baumgarten, Marjorie. "Earth and the American Dream." 17 June 1994. www.austinchronicle.com/calendar/film/1994-06-17/earth-and-the-american-dream/. Web. July 2014.

Blackburn, Steve. "American Experience: Yosemite—The Fate of Heaven (1990)." November 2013. www.alibris.com/moviesearch?qsort=&page=1&matches=0&browse=1&mtype=V&upc=0012236893233&full=1. Web. July 2014.

Brahic, Catherine. "Al Gore's *An Inconvenient Truth*: Unscientific?" *New Scientist Blogs*, 1 October 2007. www.newscientist.com/blog/environment/2007/10/al-gores-inconvenient-truth.html. Web. November 2013.

Burnett, H. Sterling. "The Truth about Al Gore's Film: *An Inconvenient Truth*." The National Center for Policy Analysis, 22 June 2006. www.ncpa.org/pub/ba561. Web. November 2013.

Cammaer, Gerda. "Praxis: Film Review: Edward Burtnysky's *Manufactured Landscapes*: The Ethics and Aesthetics of Creating Moving Still Images and Stilling Moving Images of Ecological Disasters." *Environmental Communication* 3:1 (2009): 121–130. Print.

Capturing the Friedmans. Dir. Andrew Jarecki. Magnolia Pictures, 2003. Film.

Carson, Rachel. *Silent Spring*. New York: Houghton Mifflin, 1962. Print.

"Charcoal People." Internet Movie Database, www.imdb.com/title/tt0245034/. Web. November 2013.

Cieply, Michael. "The Films Are Green, but Is Sundance?" *New York Times*, 16 January 2009. www.nytimes.com/2009/01/17/movies/17green.html?_r=0. Web. November 2013.

The Cove. Dir. Louie Psihoyos. Oceanic Preservation Society, 2009. Film.

Crude. Dir. Joe Berlinger. First Run Features, 2009. Film.

de Valck, Marijke. *Film Festivals: From European Geopolitics to Global Cinephilia*. Amsterdam: Amsterdam University Press, 2007. Print.

DC Environmental Film Festival. "About Us." www.dcenvironmentalfilmfest.org/about. 1 June 2013. Web. November 2013.

Dunn, Colin. "TreeHugger Picks: Required Reading." Treehugger, 11 December 2006. www.treehugger.com/culture/treehugger-picks-required-reading.html. Web. November 2013.

Food, Inc. Dir. Robert Kenner. Magnolia Pictures, 2008. Film.

Fox, Josh. "Affirming *Gasland*." GaslandtheMovie.com, 1 July 2010. www.gaslandthemovie.com/whats-fracking/affirming-gasland. Web. November 2013.

Gasland. Dir. Josh Fox. New Video Group, 2010. Film.

"*Gasland* Debunked." Energy In Depth, 1 July 2010. www.energyindepth.org/gasland-debunked/. Web. November 2013.

Glass, Philip. "*Anima Mundi*." November 2013. www.philipglass.com/music/films/anima_mundi.php Web. July 2014.

Grierson, John. *Grierson on Documentary*. Ed. Forsyth Hardy. London: Faber, 1966. Print.

Hansen, James "Huge Sea Level Rises A Coming—Unless We Act Now," *New Scientist*, 25 July 2007. www.newscientist.com/article/mg19526141.600-huge-sea-level-rises-are-coming—unless-we-act-now.html. Web. November 2013.

Herman, Judith. *Trauma and Recovery.* New York: Basic Books, 1992. Print.

Holden, Stephen. "Film Review: A Spiritual Trek by Yak Caravan." *New York Times*, 22 July 1998. www.nytimes.com/1998/07/22/movies/film-review-a-spiritual-trek-by-yak-caravan.html. Web. November 2013.

Holmes, E. Burton. "Mr. Holmes at Daly's Theatre." *New York Times*, 15 March 1898. Print.

Horak, Jan-Christopher. "Wildlife Documentaries: From Classical Forms to Reality TV." *Film History* 18:4 (2006): 459–475. Print.

"Hydrologic/Geomorphic Assessments," United States Environmental Protection Agency, 22 March 2013. http://water.epa.gov/polwaste/nps/watershed/hydrologic_geomorphic.cfm. Web. November 2013.

An Inconvenient Truth. Dir. Davis Guggenheim. Lawrence Bender Productions, 2006. Film.

Iraq for Sale: The War Profiteers. Dir. Robert Greenwald. Brave New Films, 2006. Film.

Johnson, Paula C. "The Social Construction of Identity in Criminal Cases: Cinema Verité and the Pedagogy of *Vincent Chin.*" *Michigan Journal of Race & Law* 1:2 (1996): 348–489. Print.

March of the Penguins. Dir. Luc Jacquet. Bonne Pioche, 2005. Film.

Marzani, Carl. *The Wounded Earth: An Environmental Survey.* Reading, MA: Young Scott Books, 1972. Print.

Mcavoy, Esme. "Who Will Pay for Amazon's 'Chernobyl'?" *The Independent*, 10 January 2010. www.independent.co.uk/environment/green-living/who-will-pay-for-amazons-chernobyl-1863284.html. Web. November 2013.

Minster, Mark. "The Rhetoric of Ascent in *An Inconvenient Truth* and *Everything's Cool.*" *Framing the World: Explorations in Ecocriticism and Film.* Ed. Paula Willoquet-Maricondi. pp. 25–42. Charlottesville: University of Virginia Press, 2010. Print.

Murray, Robin and Joseph Heumann. "Al Gore's *An Inconvenient Truth* and Its Skeptics: A Case of Environmental Nostalgia." *Jump Cut,* no. 49 (2007): n.p. Web.

Nichols, Bill. "Historical Consciousness and the Viewer." *The Persistence of History.* Ed. Vivian Sobchack. pp. 55–68. New York: Routledge, 1996. Print.

Nichols, Bill. *Representing Reality.* Bloomington: Indiana University Press, 1991. Print.

"A Place in the Land." Billings Farm & Museum, 2 July 2013. www.billingsfarm.org/exhibits-collections/film.html. Web. November 2013.

Romney, Jonathan. "What's Up, Doc?" *The Guardian*, 19 August 1997.

Seymour, Julia A. "Executive Summary: Science Fiction: 5 Years After: Networks Celebrate Al Gore's 'Inconvenient Truth,' Ignore Scientific Flaws, Criticism." Business and Media Center Institute, 23 May 2011. www.mrc.org/bmi/reports/2011/Executive_Summary_Science_Fiction.html. Web. November 2013.

Spence, Louise and Vinicius Navarro. *Crafting Truth: Documentary Form and Meaning.* New Brunswick, NJ: Rutgers University, 2011. Print.

"Stars, Twerking, Obamacare Drive Internet in 2013." Yahoo.com, 3 December 2013. https://sg.news.yahoo.com/stars-twerking-obamacare-drive-internet-2013-092953470.html. Web. July 2014.

Stone, Flo. "The EMA Awards." Environmental Media Association. www.ema-online.org/awards. Web. November 2013.

Stone, Flo, to Charles Musser, telephone conversation, 15 March 2010.

Stringer, Julian. *Regarding Film Festivals*. Bloomington, Indiana University Press, 2003. Print.

Tucker, Ken. "TV Review." *Entertainment Weekly*, 23 April 1993. www.ew.com/ew/article/0,,306326,00.html. Web. November 2013.

Turan, Kenneth. *Sundance to Sarajevo: Film Festivals and the World They Made*. Berkeley: University of California Press, 2003. Print.

Vampires, Devilbirds and Spirits: Tales of the Calypso Isles. Dir. Nick Upton. British Broadcasting Corporation, 1994. Film.

"Vision Man: An Eskimo Hunter" Academic Video Store. www.academicvideostore.com/video/vision-man-eskimo-hunter. Web. July 2014

Who Killed the Electric Car? Dir. Chris Paine. Electric Entertainment, 2006. Film.

Wildlife Film Festival. http://wildlifefilms.org/about/. Web. November 2013.

Winston, Brian. "The Tradition of the Victim in Griersonian Documentary." *New Challenges in Documentary*. Ed. Alan Rosenthal. pp. 269–287. Berkeley: University of California Press, 1988. Print.

Young, Andrew. "*Spirit of Kuna Yala*." 1 November 2001. www.imdb.com/title/tt0926223/. Web. July 2014.

3 Great Southern Wounds

The Trauma of Australian Cinema

Mark Steven

Lindy and Michael Chamberlain's first daughter, Azaria, was born on June 11, 1980. When Azaria reached two months of age, Lindy and Michael took all three of their children on a camping trip to Australia's Red Center—to the sandstone megalith then named Ayers Rock. One night into their stay, on August 17, Azaria vanished. Although a search party scoured the desert, all they found were the bloodstained remains of an infant's clothing. The Chamberlains insisted that a dingo ate their baby; the judiciary disagreed, prosecuting Lindy and Michael as her murderers. Six years later and near the site of Azaria's disappearance, an English tourist, David Brett, fell to his death while attempting to scale Ayers Rock. His body was found eight days later in an area crowded with dingo lairs. As police searched for missing bones that might have been carried off by the native animals, they also uncovered an unrelated piece of evidence: the tattered remnants of Azaria's matinee jacket. The Chamberlains' lawyers launched an appeal based on this newfound evidence and, consequently, Lindy and Michael were acquitted of their daughter's murder. The inconclusive findings of a subsequent inquest were released on December 13, 1995, wherein the coroner's report is deeply unsettling:

> Azaria Chantel Loren Chamberlain died at Ayers Rock on 17 August 1980. As to the cause of her death and the manner in which she died the evidence adduced does not enable me to say. I therefore return an open finding and record the cause and manner of death as unknown.
>
> (Lowndes n.p.)

From whichever angle you look at it—whether she was devoured by native fauna or butchered by her parents—Azaria Chamberlain now reaches out as the paragon of familial trauma whereas that mythic dingo remains the avatar of a particular breed of ecological violence. Although the event eludes narrative definition, the "cause and manner of death unknown," when brought together in the shadow of Ayers Rock, Azaria and her dingo metonymically embody the structural realization of Australia's own "eco-trauma."

I want to begin by suggesting that the denotations of "eco" and "trauma" are ultimately unrelated and can only be combined through disjunctive synthesis—that the combinatory dash securing the two as "eco-trauma" carries enough conceptual weight to warrant some discussion of its own. Bearing in mind that "eco-" broadly refers to the relations of organisms to one another and to their physical environment, perhaps Catherine Malabou's neurological revisions of the psychoanalytic enterprise might prove helpful in linking ecosystems to a theory of trauma. To advance a definition, then, "trauma" refers to the psychological intrusion of something radically unexpected and ecologically exterior—something which cannot be assimilated into the psyche without returning in neurotic symptoms or, as Malabou suggests, something which cannot be assimilated at all. It was Jacques Lacan who argued (after Sigmund Freud) that traumatic events repressed in the symbolic order will always return in the real; and, in *Les nouveaux blessés* (2007), Malabou reviews this theory in light of "the delicate echoes" between the internal and an external real, between the psychological mind and the neurological brain (see Žižek 2010). According to Lacan, the "real" comprises ecological and psychological phenomena that defy presentation and, for him, encounters with the external real derive their properly traumatic impact from the way they touch on a preexisting, preconscious "psychic reality"—it is thus that the real of ecological violence should transform itself into traumatic symptoms, by stimulating the already established psychological real (see Lacan 1977).

Given the present narrative of parents, babies, and dingoes, any psychoanalyst worth his or her salt might very well struggle to think about Azaria outside of that structurally familial and highly symbolic frame, perhaps even implanting her parents' encounter with a brutal exterior into the symptomatic dream sequence of its interior counterpart. "Father," an unbearable fantasy of the dead child, would reproach the traumatized patriarch: "Can't you see I'm being eaten by a dingo?" Here the violent intrusion of an ecological exterior owes its traumatic effect to a resonance with familial anxiety; an exemplary ecological shock, it would seem, awakens a set of psychological preconditions. This is all very fine and would make for interesting analysis, but we ought to sharpen these ideas on Malabou's basic reproach, that Lacan and Freud never fully accepted the neurologically and thus purely destructive power of external shocks—that ecological violence might simply destroy the psyche of its victim without awakening any inner truth. My sense is that a specifically Australian sense of eco-trauma is sourced not only in unconsciously determined anxieties but, rather and also primarily, in the symbolically indeterminate shocks which consciousness can in no way appropriate or integrate. An infant vanished into the external real and so an ecological threat was magnified; internally, the event lacks definition and the wound is open, un-cauterized by trauma—all of which yields the question: How do irreconcilably violent phenomena penetrate the psyche and calcify as trauma?

Alain Badiou provides an invaluable deconstruction of the symbolic inde-
terminacy ascribed here to ecological violence as well as to the preconditions
of the psyche. "Nothing can attest that the real is real," he writes, "nothing
but the system of fictions wherein it plays the role of the real" (Badiou 49).
Our insatiable desire to appropriate or integrate the shock of an ecological
exterior is only ever manifest in its representation, Badiou would argue,
in the source of that shock's own veridical semblance. "Beyond semblance
there is a *necessity* of semblance," he continues, "which has perhaps always
constituted the real" (49). If the real can only present itself in representa-
tional semblance, I should like to conclude these opening remarks with two
interrelated hypotheses: first, that cinema stands as foremost medium of the
real—of screens, surface, and the spectacle—and, second, that Australia's
ontologically constitutive "system of fictions" comprises its film industry,
whose output provides traumatic short circuits between the internal and the
external real, which together light up a national sense of eco-trauma. To be
sure, any cross-resonance between "eco" and "trauma" must be mediated
so as to deaden the impact of the former, somewhat ironically, by intensify-
ing its traumatic reverberation through the latter. While cinema furnishes
the devastating impact of external violence with some degree of symbolic
efficacy, by dint of an imposed familiarity, emphasis in cinematic representation simultaneously shifts from destructive events to their traumatic reckoning. With little surprise, then, did the Chamberlain docudrama *A Cry in the
Dark* (1988), centre on the temperance of grief itself, only to resolve in narrative when Lindy (Meryl Streep) abandons her stoicism for the semblance
of a mother's heartache.

Australia has its own native strains of ecological violence as well as its
own movie industry, and between the two that task of generating familiarity
is elevated to the productive power of a trauma industry. The efflorescence
and influence of this industry could be argued to have shaped Australian
cinema more generally, wherein forcibly familiar inflections crystallize into
tropes which accumulate as genres. This chapter is wagered on that very
proposition, that Australian cinema is linchpin to Australia's sense of eco-
trauma, and that thinking about eco-trauma will account for some of the
eccentricities in Australian cinema. An ancillary argument might propose
that anything like the "uncanny" in Australian cinema, or at least in recent
films, might have something do with the production of eco-trauma and its
disjunctive synthesis of ecological violence and the psychologically familiar.
While Ken Gelder and Jane Jacobs's tremendous book, *Uncanny Australia*
(1998), explains such phenomena from a postcolonial perspective and in
response to the displacement of Australia's Indigenous population, here it
will only be added that a theory of eco-trauma has something to say about
the cinematic realization of that very displacement, an idea with which this
chapter concludes. What follows, then, are four discussions of popular Australian cinema, all of which consider films that mediate between ecological
violence and psychological trauma, between the external real and its internally metastasized symptoms.

VIOLENT TOPOGRAPHIES

To think about Australia as an ecosystem requires at least some consideration of its ecologically conceived elements, merging into heterogeneous environments that are aggregated together with an idea of place. As with any other place, Australia is, on the one hand, a physical entity made up of unstable ecosystems and, on the other, an assemblage of equally volatile representations relating landmass to consciousness. Various disciplines teach that ecological Australia's amorphous instability has found form in evolutionary habituation, generating antipodean adaptations to the arid Australian landscape. Here flora and fauna have acclimatized to those multiple destructive forces—droughts, fires, floods, and so on—whose presence and frequencies are distinctly Australian. If a theory of eco-trauma takes the psyche as a traumatized, internal adaptation to its violent, external topography, and if cinema is a medium for translating between the two, then how, exactly, does that translation operate? These first three examples demonstrate how cinema actively re-routes violent fantasies about generically conceived Australian topographies through forcibly familial and accordingly familiar psychological territory, amplifying by way of transference any resonance between ecological violence and preexisting psychical realities, thus giving shape to eco-trauma.

The underrated horror film *Long Weekend* (1978) is one of the best celluloid offerings of violence originating in Australia's natural environment. Unlike the myriad other films taking on a similar theme, this thriller avoids capturing a singular menace and instead fantasizes a multitude of threats that could collectively comprise the entirety of an Australian ecosystem. After an extended long shot of the curving South Australian shoreline followed by a juxtaposing cut to an angular urban sprawl, the camera's omniscience betrays its two main subjects, Peter and Marcia, a suburban couple who are, in her words and at that time inexplicably, "not talking at the moment." Psychoanalysis would suggest that their dire relationship speaks to the pitfalls of sexual difference as amplified by hidden trauma and, when the disconnected couple goes camping, the strained energy of that differential trauma plays out on the ecosystem with the revelatory charge of a fully fledged transference. From the moment Peter and Marcia hit the road, they seem intent on eradicating as much of the natural environment as stumbles into their headlights. After an unblinking and daylong massacre of a South Australian menagerie, Peter—having perhaps worked something through—drunkenly tries to force a sexual connection with Marcia. "Sorry," she pushes him away. "It's alright," he tells her. "How long did the doctor say? Weeks? Months?" Soon into their stay, the imperceptible real of a beachfront campsite appears to mobilize in a spectacle of nature's fury, as though the ecosystem is rearing up to destroy the intruders and resume quiescence. This is ecological violence at its symbolically indeterminate, Darwinian best—and yet, as the couple begins to fear for lives and safety, narrative redirects the violence through that hidden trauma blocking their sexual relation. "Screwing the neighbours," he accuses her, "and murdering the unborn." Marcia was unfaithful and had an abortion, and this pre-existing trauma re-emerges in transference when she gruesomely shatters the near-hatched egg of a sea eagle; and, in counter-transference, when the whole of that ecological space fills with an omnipresent shrieking not dissimilar to that of an infant. "I can hear her crying," Marcia bleats into the setting sun, doubling the howls of native fauna with those of her dead child. To compress the final half-hour down and into a statement on the realization of eco-trauma, nature's assumed retaliation acquires an uncanny similitude to the psychological symptoms of Peter and Marcia's already traumatic relationship: Each cannot relate to another, together neither can relate to the natural other, and these non-relations are ultimately destructive, leading to both of their deaths—not through the external ecosystem alone but at the hands of psychologically reflexive humans embedded within that system, with Peter inadvertently spearing Marcia through the throat and a truck running Peter down, dragging his wet viscera along dry bitumen.

Shifting from the natural to an emphatically urban ecosystem, Rolf de Heer's *Bad Boy Bubby* (1993) presents itself as one of the strangest films set in any Australian city. This pitch-black comedy's claustrophobic opening scenes introduce a heavy-handed Oedipal narrative replete with uncomfortably incestuous heavy petting. Bubby, a thirty-five-year-old man, lives in a filthy, grey apartment where he maintains a torturous, sexual relationship with Flo, his mother. Bubby has never ventured beyond the apartment on fear of death—his mother has convinced him that outside is poisonous gas (Figure 3.1). But, when she brings home a priest who claims to have seeded Bubby, the Oedipal narrative strikes its zenith and then crosses another taboo, with Bubby cling-wrap suffocating his father and mother, whose corpse he takes to bed and tries to awaken. This is all traumatic stuff, but it has little to do with ecology—save for the fantasy of poison; however, once Bubby leaves his flat, the ecological violence in Adelaide, capital city of South Australia, is coupled with a categorically unresolved Oedipal narrative so as to render that disjunctive impact. Beyond the threshold to his flat, Bubby comes up hard against Adelaide's local fauna. "Get off the road you greasy mongrel," one man drunkenly shouts from the back of a truck, inducting Bubby into the urban ecosystem.

In the subsequent twenty-four hours, Bubby witnesses a robbery and is attacked by a dog, thrown out of a shop, beaten by a police-officer, threatened with a chainsaw, and finally co-opted into robbery, arrested, and raped while in lockup. There is no poisonous gas flooding the Adelaide airspace, but the city is undoubtedly toxic and potentially fatal. Perhaps the most memorable quality of this film is Bubby's diction—that it is limited to mimetic repetition of what has been said to him in oftentimes violent or traumatic circumstances, and that Bubby thus internalizes and rearticulates the soundscape of Adelaide, refracting sonic traces of ecological violence through his own damaged psyche, stretching an ecosystem over the contours of superannuated desire. A nihilistic preacher tells him that "there is no God" and that "we

Figure 3.1 Director Peter Weir emphasizes in his framing the magnitude of the natural world that tragically engulfs the students in *Picnic at Hanging Rock*.

arrange more order and harmony." Order through harmony, he ought to have said, as Bubby reorganizes the ecological violence of an unruly and amoral cityscape into the most harmonious of forms: Bubby, by hapless chance, finds himself fronting a rock band. Donning his father's clerical robes and assuming that paternal identity, Bubby's blistering performances comprise of his belting out in broken baritone a sonic bricolage of ecological violence and internal traumas, fusing the two as eco-trauma. The narrative resolves when Bubby transcends the baleful violence of Adelaide's urban sprawl hand-in-hand with Angel, a sexual surrogate for Bubby's maternal love-object. "Like mom," Bubby described her breasts. In the final shots, the camera pulls away from Bubby, who is here playing with his own two children on a deracinated island of grass at the heart of an urban wasteland; it is an echo of his flat, complete with children, but Bubby has assumed paternal order and with that he is psychologically fortified against ecological violence.

To look at any map of Australia will confirm that between natural environments and pulsating cities run national highways—vast, interstitial non-places, depicted nowhere more electrifyingly than in the original *Mad Max* (1979). "High Fatality Zone," reads one road sign in the opening scene, a cenotaph to eco-trauma, with a skull and crossed bones painted on another. Here bands of "terminally psychotic" bikers administer meaningless brutality, destroying all that is in their path. That this violence is as meaningless as the external threats of *Long Weekend* and *Bad Boy Bubby* is made evident from the film's outset, when it is metonymically compressed into the self-assumed leader of the highway gangs. The Nightrider refers to

himself as a "fuel-injected suicide-machine," a nitrous-guzzling personification of the psychical death-drive, and is run off the road into his own
fiery end by leather-bound highway cop Max (Mel Gibson). Max's desire
to maintain some semblance of order on the roads is a desire soon conflated with his own paternal dilemmas, for in this dystopic environment
Max and his wife, Jessie, are trying to raise a child. "I'm scared," Max
tells his superior. "It's a rat circus out there, and I'm beginning to enjoy it.
Look, any longer out on that road and I'm one of them, a terminal psychotic, except that I've got this bronze badge that says that I'm one of the
good guys." "Take a vacation," Max is told, and there, on vacation, both
narratives converge, with meaningless violence cleaving through familial
order. In the film's most devastating scene, mounted bikers pursue Max's
wife and son down the highway, with the mother on foot and carrying her
child. Two speeding-shot reverse shots are coupled with the thunderous
cacophony of bikes' approach—from the bikers' perspective and then a
returned, over-the-shoulder gaze from Jessie, back to the biker, then down
to his odometer, indicating further acceleration—and, with all the camera's
kinetic energy behind them, the bikes plow into their victims, killing both.
Only after this violence befalls his family does Max take to the roads and
against the bikers with any serious intent, and it is thus that the narrative
of *Mad Max* requires external horror to resonate with internal trauma so as
to bring home the transference of an eco-trauma. That Max murders each
and every member of the gang, one by one, and by increasingly elaborate,
sadistic, and vengeful methods, speaks to his working through paternal
failure with all the reprisal of violent transference. Max fulfils his own
prophecy, becoming "one of them," combining their ecological violence
and his psychological trauma, internalizing the road while imposing on it
his own inner disquiet. As one biker characterizes a car's wreckage early
in the film, while listening to an acne-faced teenager describe it as having
been "chewed up and spat out"—"Perhaps," he mutters, "it was the result
of an anxiety."

The recasting of ecological violence through anxious energy seems to be
a productive gesture shared across all three films. Here I want to emphasize
that when it comes to cinema, the topographical violence of a microcosmic
ecosystem only develops into full-blooded trauma once routed through pre-
existing psychological paradigms. By refracting distinctly Australian violence through psychological preconditions, forcing connections between the
internal and the external real, the psyche is itself positioned as a product
of that evolutionary habituation, conflating external shock with internal
anxiety so as to make sense of the former, awaken the latter, and to thereby
deliver eco-trauma. One of the major outcomes to this cinematic production
is a specifically Australian idea of character, stretched between narrative and
topography, crystallized through psychology, and (as we shall now see) split
along the lines of sexual difference.

THEY'RE A WEIRD MOB

The historical fact that Australia's first feature film was about a gang of out-law bushrangers scarcely requires mentioning. Although there is no complete print left of *The Story of the Kelly Gang* (1906), commentaries and a surviv-ing eleven minutes give the impression that cinema's national inauguration was not just formally seminal but also thematically influential. Significant here is the relation between a singular man and the masculine collective—tension between the gang's leader, Ned, and his lesser-known brethren—when that collective is driven to an extreme self-realization at the explosive crossroads of economic subjugation and ecological violence. Since *The Story of the Kelly Gang*, Australian cinema has time and again taken as its sub-ject the masculine character born of that confluence, as manifest in famil-iar figures like Crocodile Dundee (Paul Hogan), Mad Dog Morgan (Dennis Hopper), and a multitude of variants on the theme of Ned Kelly (Mick Jagger, Yahoo Serious, Heath Ledger)—all of which have amalgamated and mutated into the vicious *ne plus ultra* of their own reckoning with the truly terrifying Mick Taylor (John Jarratt) of *Wolf Creek* (2005). However, and withdrawing from the fantastical exemplification embodied by any of these mythic figures, the masculine character is more often than not forged in the collective—in tightly bunched pockets of homosocial masculinity wherein economic and then ecological violence are returned in toxic personalities. When it comes to these oftentimes sexist, racist, and ultimately xenophobic cloisters, the idea of eco-trauma might provide some answers to the question of what it means to be an Australian man with other Australian men.

Wake in Fright (1971)—also known as *Outback*—is a film about a young man's fall from grace, as he devolves from the respectable school-teacher in one outback town into a suicidal, kangaroo-butchering drunk-ard in another. John Grant is, by his own admission, a "bonded slave to the education system," forced to teach in uttermost rural Australia, in the cramped schoolhouse of Tiboonda (actual location is Broken Hill, New South Wales), and, as the narrative opens, he is homeward bound to Syd-ney, where he plans to spend six weeks on the beach with his girlfriend. Before flying back to the coast, circumstances require he spend one night in a town he evidently and openly finds detestable, Bundanyabba, the pri-mary virtue of which, a cabdriver tells him, is that it is "a friendly place, the Yabba." This could easily translate into the Yabba's actualization as a homosocial pressure cooker, the central base to an army of outback mine-workers, where individuals are forced together by their unforgiving sur-roundings and, due to an undrinkable water supply dovetailing with an observance to local custom, are bound by the prodigious consumption of alcohol. In the Yabba, beer serves as liquid currency to an economy of rural masculinity. En route to the Yabba Grant declines a beer from his fellow travellers and instead opts to fantasize about his girlfriend, therein conjuring an oceanic space of breaking waves before which he strokes her chest with (of course) an open beer bottle. Once arrived, however, Grant is forcibly lured into buying rounds with the Yabba's police officer, Jock Crawford (Chips Rafferty), and, with critical capacities well and truly battered, he gambles away his paycheck and flight money in a game of Two Up. While the game itself speaks to an idea of ecological violence manifest in homosociality—the chance spin of two coins wields power to decimate one's income, sanctified by the rule-bound relation one shares with fellow players—here I want to stress that, through the combinatory allure of beer and brethren, partners united against a pitiless outback ecosystem, Grant traps himself in the Yabba, unable to fly home.

Two scenes exemplify *Wake in Fright*'s articulation of eco-trauma. After being invited home with a bandy-legged man, Tim Hynes, who hosts three other men for a drinking session, Grant wanders outside with Hynes's daughter, Janette, and half-heartedly tries to seduce her. "What's the matter with him," asks one of the men inside, "rather talk to a woman than drink?" Janette throws herself down in front of Grant and, writhing about in the dust and dirt, pulls him on top of her. In one of the more awkward and abject sexual encounters to grace Australian celluloid, Grant kisses her, and then proceeds to vomit for an excruciating half-minute. After Grant's "episode with Janette," he returns to the party and sinks enough beer to lose consciousness, but not before bragging that he once won a silver medal for target shooting. With this information and well into the bleary-eyed day after, three of the men—Doc, Dick, and Joe—take Grant kangaroo hunting. Having gunned down and run over several of the animals, Joe finishes off a still-kicking animal with the blade of his knife. "Then he rips his guts out with his hind legs," Doc explains the kangaroo's defence mechanism, conjuring the return of an ever-immanent ecological violence. Inebriated, Grant tries the same with a smaller and equally wounded animal—all the time it kicking and clawing at him—and together he and it roll about in the dirt. Rather than slash its throat he plunges the knife in again and again and to the applause of the other men. This scene, with its appalling redolence of Grant's night with Janette, marks the sublimation of sexual trauma into the bonds of masculinity via beer-fuelled carnage.

Psychological violence is thus transferred into ecological violence and both are conflated, or even compounded, by the figure of Doc, who tells John that Janette and the kangaroo were equally slow-moving targets—and who amplifies the homosocial bond to the intimate power of a homosexual relation, pouring beer over a shirtless Grant and injecting eco-trauma (as it were) via a forcefully implied act of sodomy. Whether sodomy ever really occurs is never made clear, but this narrative veil brings an end to all of Grant's nights in the Yabba: The scenes finish with twenty-four frame flashes, often tilting and panning, to blackout entirely in drunken anamorphosis. The film's three blackout scenes are followed by cinematically rendered hangovers, wherein everything is overexposed and punishingly loud, replete with the surround-sound drilling of mosquitoes, and Grant's memories are only ever available for speculation. It is thus that *Wake in Fright*'s liquid economy is concomitant with the production of eco-trauma: Beer alleviates ecological violence but also fuels and thus compounds it into an experiential though repressible trauma. Hence, in a reversal of the opening train ride, with Grant now travelling from the Yabba and back to Tiboonda, he happily accepts a beer from his fellow travellers, as though to complete his apprenticeship in Australian masculinity.

This eco-traumatic idea of camaraderie against circumstances as enabled by drinking is a common theme in Australian cinema. It finds commensurately traumatic form in *Romper Stomper* (1992), a film whose narrative is driven by an explosively xenophobic reaction to Vietnamese immigrants buying out a Melbourne bar, but it is also apparent in arguably less traumatic films about the Australian workforce such as *Sunday Too Far Away* (1975) and Michael Powell's *They're a Weird Mob* (1975). The postindustrial landscape of Newcastle, New South Wales, has more than once played backdrop to the effects of this juncture, in *Bootmen* (2000) and *Newcastle* (2008). And, taking that cultural conjunction to a death-ride highway is the harrowing road film *Stone* (1978).

VANISHING POINTS

If Australian cinema's moment of inception was with *The Story of the Kelly Gang*, the earliest feature of which all parts still survive, *The Romance of Margaret Catchpole* (1911), has proven just as influential, shifting attention from the masculine so as to present a romantic fantasy of what it meant to be a woman at the turn of the century in Australia. There is very little to be said about ecological violence here, but worth mentioning is that Catchpole is only deported to Australia, as a convict, because she expropriated a horse to assist her lover's jailbreak. From this point onward and until recently, women have occupied a curious place in Australian cinema. Unlike men, who are forced together by and against threats of the ecological, and who recode that violence as conditional to an inebriated, masculine sense of character, women are forcefully put under erasure; they are obliterated from or assimilated into the ecosystem altogether, but only after their gender is performed as spectacular fantasy.

Peter Weir's *Picnic at Hanging Rock* (1975) generates what is, perhaps, the most celebrated eco-trauma in the whole of Australian cinema. Its opening title card reads,

> On Saturday 14th of February 1900 a party of schoolgirls from Appleyard College picnicked at Hanging Rock Mt. Macedon in the state of Victoria. During the afternoon several members of the party disappeared without trace . . .

That ellipsis carries an unresolved narrative out of written syntax and onto several long, hazy shots of a eucalypti covered hill and, after the camera reemerges from tall grass, Appleyard College comes into focus, obscured by encroaching trees (Figure 3.2). The narrative commences on Saint Valentine's Day and from the outset implies lesbian desire, with the girls reciting love lyrics penned in one another's honour. "You must learn to love someone else, apart from me," Miranda says to Sarah. "I won't be here much longer." The school's headmistress furnishes Miranda's prescient evocations with a multiform ecological threat. "Once again, let me remind you, the Rock itself is extremely dangerous," she says. "Therefore, forbidden is any tomboy foolishness in the matter of exploration, even on the lower slopes. I also wish to remind you, the vicinity is renowned for its venomous snakes and poisonous ants of various species." Between Miranda's prophecy and the headmistress' warning, the possibility of ecological violence is raised to probability, a Damoclean threat poised above this fantastical vision of a fragile femininity.

The scenes at Hanging Rock have an eerie and almost dreamlike quality to them. That some sort of ecological violence is coming to fruition seems implicit when Miranda unlatches the gate onto the Hanging Rock parkland. Here, and centring on Miranda, a curious set of shots fade into and out of one another, all of which overlay phenomena surrounding her: a flock of lorikeets flying in a gyre, a forest of eucalypti, three black horses,

Figure 3.2 Bubby lives in squalor and spars with his mother, yet is convinced the world outside his apartment is toxic in *Bad Boy Bubby*.

kicking and galloping as though spooked, and the haunted expression of Miranda's own sky-turned face. Each shot lingers over the others before fading into the next, giving the appearance that the images are assimilating one another into unseen celluloid depths, before resuming the narrative with a vertiginous low shot, looking up at the girls who now stand in the foreground of Hanging Rock. Miranda emerges, right arm raised and kitchen-knife in fist, before bringing the blade down into an as yet unseen Saint Valentine's cake. Three girls are granted permission to hike up the base of the Rock and a fourth, Edith, follows. Last to leave, Miranda turns and waves to her teacher—she is framed in a fantastically faraway soft focus and waves in slow motion. "Now I know," says her teacher, looking down to a book of artworks. "I know that Miranda is a Botticelli angel." As the four girls ascend Hanging Rock, the camera's perspective seems to splinter on its topography; their ascent is shot from various cracks and crevices, from high and low, and with inconsistent proximity. Of the four, the three who initially set out are captured by a montage of cross-fades; they remove their stockings and shoes, cease speaking, and climb between two boulders and beyond the threshold of perceptibility—beyond a perspective that belongs to Edith, who only ever trails behind. Edith begins to scream, a momentary jump cut shows Hanging Rock shrouded in red mist, and Edith is filmed running down the hill, still screaming, an image captured from a moving, aerial camera. Some manner of ecological violence has befallen the girls: They have, it seems, been assimilated into the face of Hanging Rock.

The girls' disappearance is never explained, and so the events at Hanging Rock remain mysterious and beyond symbolic purchase. However, the girls' final encounter with other humans is worth consideration. As they hop across a river two boys spy them and one comments, "Thought the little fat one was gonna take a bath," he says. "Have a look at the shape of the dark one with the curls. Built like an hourglass. And have a go at the last one, the blonde. She'd have a decent pair of legs, all the way up to her bum." The three corseted girls and especially Miranda are made objects of the male gaze all the more literally than by the camera's already voyeuristic operations. Their femininity is sexualized to the point of fantasy and that sexualization comes to bear on their curious behaviour minutes before they vanish. When Miranda removes her stockings, the camera presses near for a fetishistic close-up of her left leg from just above the knee as her fingers slowly slide a black stocking down to her ankle, revealing but also stroking pale skin—a self-consciously sensuous gesture. This shot recalls the boy's comment—"all the way up to her bum"—and so Miranda, in the preparatory moment of her disappearance, undergoes one final metamorphosis into an exemplary fantasy of feminine sexuality.

A similar logic is at work in equally if not more traumatic films that operate without the overt threat of ecological violence. In *Puberty Blues* (1981), for instance, two teenage girls subject themselves to the sexual tortures of a band of surfers so as not to fade into social obscurity. More recently, however, femininity has found similar destination in the comparably superb films *Somersault* (2004) and *Beautiful Kate* (2009), both of which revise the idea of a mythic vanishing point. Yet nowhere is the echo of *Hanging Rock* heard louder than in the experimental verisimilitude of *Lake Mungo* (2008), a fictional documentary about a sixteen-year-old girl named Alice who foresaw her own death in the Australian outback and whose spectral apparition returns first in dreams but then in the swarming pixels that comprise the photography and videos made by her traumatized family. "Everything begins and ends at exactly the right time and place," says Miranda before vanishing, yet Alice's time is out of joint—through the looking glass she encounters the nightmarish prolepsis of her own ecologically violent death—and it is thus that Alice and Miranda both exemplify the feminine character of eco-trauma. In these examples, at least, women are not about to experience the trauma of ecological violence; they are, rather, always already victim to that violence, and so their assimilation, their passing through a vanishing point, opens up an ultimately gendered manifestation of eco-trauma.

TRAUMATIC DREAMING

From the very beginning, with *The Story of the Kelly Gang*, films produced on Australian soil have collectively forged paradigmatic characters out that tension between ecological violence and psychological trauma, from the maternal anxieties of *Long Weekend*, through the ferocious paternity of *Mad Max*, via urban maturation in *Bad Boy Bubby*, outback masculinity in *Wake in Fright*, and the vanishing point of *Hanging Rock*: If all of this is not the cinema of eco-trauma, it can only be the celluloid register of an explosively unstable ecosystem; and if the cinema of eco-trauma organizes ecological violence into preexisting psychological narratives, there is an ethical charge to be delivered on what that procedure might achieve when operating in reverse. The following paragraphs comprise a provisional reflection on a particular kind of ideological mystification, when cinema represents the categorically organized conditions of trauma as though they were only ever eruptions of ecological violence.

Any discussion of trauma in Australian cinema will sooner or later arrive at the question of Indigenous history and its transposition onto film. The most traumatic event faced by Australia's Indigenous population is the deliberate act of protracted genocide, inflicted by white Australians over the course of decades, in a program whose implementation was far from ecological and is, rather, loaded with ideological significance. However, the narrative techniques for producing eco-trauma have been hijacked to mystify this very significance, thereby disposing of engagement with the traumatic actualities of Australian history. While it is true that films depicting the annihilation of Australia's Indigenous population are variously traumatic—consider, for instance, the melancholia and melodrama of *Tracker* (2002), *Rabbit-Proof Fence* (2002), and, looking back somewhat, *Walkabout* (1971) and *The Chant of Jimmie Blacksmith* (1978)—most of these films re-route calculated, genocidal violence back through the sentimental psyche and, furthermore, transpose that traumatic energy into the rendering of Australia's ecologically violent topographies. On February 13, 2008, Australian Prime Minister Kevin Rudd issued a long-awaited apology to the Indigenous population of Australia, within which he said,

> That is why the parliament is today here assembled: to deal with this unfinished business of the nation, to remove a great stain from the nation's soul and, in a true spirit of reconciliation, to open a new chapter in the history of this great land, Australia.
>
> (Rudd n.p.)

These lines speak to the persistent guilt of a land-tied trauma—to a "great stain" on the "soul" of Australia—and what I should like to add for this chapter's conclusion is that Rudd's speech exemplifies an anxiety particular to Australian cinema, an anxiety that has everything to do with colonial projects and their means of generating stain-free, ethnically cleansed topographies. This anxiety is incommensurately manifest across two films whose comparison should call into question their ideological divergence.

In Baz Luhrmann's bizarrely problematic *Australia* (2008), a boy of mixed heritage called Nullah attains narrative significance due to his serendipitous proximity to Lady Sarah Ashley (Nicole Kidman), a British aristocrat who develops maternal designs on his adoption. When Nullah leaves Lady Sarah's farm to go walkabout, she succumbs to the act of a grieving mother. The twisted irony should be preserved, that Lady Sarah's pain is only a culturally ignorant simulation of that which ought to have been delivered, if at all, by the boy's actual mother had she not died trying to prevent his being taken away and in all likelihood assimilated into a white family. That loss is, to the surrogate mother, without cause—it is the ecological force of Aboriginal mysticism translated, for the viewer, into Lady Sarah's own psychological trauma—and not the result of a political, colonial, and genocidal sequence. After all, *Australia* revels in its own fantasy: Its visual topographies quite literally conform to spectacular maps that are superimposed over the real landscape they almost always subsume, and upon which *Australia*'s glossy violence is weighed down by so much symbolic freight as to sink under the weight of its own aggressive sentimentality. When, in the opening scene, an Aboriginal spear (tossed by an unseen white man) skewers Lady Sarah's husband, this stylized image lacks any systemic purchase; its trauma ultimately preserves Lady Sarah's familial fantasy and opens up a narrative passage for her to project that onto the outback. However, another film, released three years earlier, inverts the traumatic efficacy of that cinematic misdeed. In an ultraviolent apotheosis to end the first act of John Hillcoat's *The Proposition* (2005), the lone rider Charlie Burns (Guy Pearce) wakes in a mire of outback dust to find his horse slaughtered, already attracting flies. He draws himself upward, pistol cocked, takes half a step, and a spear plunges through his chest, impaling him, standing him up like one arm of a bloodied pair of compasses. He spits gore: a living calliper, pouring ichor onto an imaginary map. When he casts his eyes sunward but to gaze on a band of Aboriginal men and before or after the off-screen shotgun blast, a man's head detonates, tearing open from right to left, from the inside out, horrifically exploding any resonance between internal and external real. The affective shock cannot be assimilated: This tremendous action-image opens up an ontological fissure in the heart of its now petrified narrative and fills it neither with the fantasy of violent topography nor with psychological trauma—off screen and filling that hole sets the malodorous coagulum of an Australian real.

WORKS CITED

Australia. Dir. Baz Luhrmann. Twentieth Century Fox, 2008. Film.
Bad Boy Bubby. Dir. Rolf de Heer. South Australia Film Corporation, 1993. Film.
Badiou, Alain. *The Century*. Cambridge, UK: Polity Press, 2007. Print.
Beautiful Kate. Dir. Rachel Ward. Roadshow Entertainment, 2009. Film.
Bootmen. Dir. Dien Perry. Fox Searchlight, 2000. Film.
The Chant of Jimmie Blacksmith. Dir. Fred Schepisi. Umbrella Entertainment, 1978. Film.
A Cry in the Dark (aka *Evil Angels*). Dir. Fred Schepisi. Cannon Group, 1988. Film.
Gelder, Ken and Jane Jacobs. *Uncanny Australia: Sacredness and Identity in a Post-colonial Nation*. Melbourne: Melbourne University Press, 1998. Print.
Lacan, Jacques. *The Seminars of Jacques Lacan. Book XI: The Four Fundamental Concepts of Psychoanalysis, 1964*. Ed. Jacques-Alain Miller; Trans. Alan Sheridan. New York: W.W. Norton and Company, 1977. Print.
Lake Mungo. Dir. Joel Anderson. Arclight Films, 2008. Film.
Long Weekend. Dir. Colin Eggleston. Australian Film Commission, 1978. Film.
Lowndes, John. "Analysis & Findings of the Third Coroner's Inquest." *Coroner's Inquest Into the Death of Azaria Chamberlain*. December, 1995. http://law2.umkc.edu/faculty/projects/ftrials/chamberlain/lowndesreport.html. Web. July 2014.
Mad Max. Dir. George Miller. Kennedy Miller Productions, 1979. Film.
Malabou, Catherine. *Les nouveaux blessés: De Freud a la neurologie: Penser les traumatismes contemporains*. Paris: Bayard, 2007. Print.
Newcastle. Dir. Dan Castle. Australian Film Finance Corporation, 2008. Film.
Picnic at Hanging Rock. Dir. Peter Weir. Picnic Productions, 1975. Film.
The Proposition. Dir. John Hillcoat. First Look Pictures, 2005. Film.
Puberty Blues. Dir. Bruce Beresford. Roadshow Australia, 1981. Film.
Rabbit-Proof Fence. Dir. Phillip Noyce. HanWay Films, 2002. Film.
The Romance of Margaret Catchpole. Dir. Raymond Longford. Spencer's Pictures, 1911.
Romper Stomper. Dir. Goeffrey Wright. Film Victoria, 1992. Film.
Rudd, Kevin. "Apology to Australia's Indigenous Peoples," February 13, 2008. Speech.
Somersault. Dir. Cate Shortland. Magnolia Pictures, 2004. Film.

Stone. Dir. Sandy Harbutt. Hedon Pictures, 1974. Film.

The Story of the Kelly Gang. Dir. Charles Tait. J & N Nevins Tait, 1906. Film.

Sunday Too Far Away. Dir. Ken Hannam. South Australian Film Corporation, 1975. Film.

They're a Weird Mob. Dir. Michael Powell. Williamson-Powell International Films, 1966. Film.

Tracker. Rolf de Heer. South Australia Film Corporation, 2002. Film.

Wake in Fright (aka *Outback*). Dir. Ted Kotcheff. Madman Entertainment, 1971. Film.

Walkabout. Dir. Nicolas Roeg. Twentieth Century Fox, 1971. Film.

Wolf Creek. Dir. Greg MacLean. FFC Australia, 2005. Film.

Žižek, Slavoj. *Living in End Times.* London and New York: Verso, 2010. Print.

4 Into the Wilde?

Art, Technologically Mediated Kinship, and the Lethal Indifference of Nature in Werner Herzog's *Grizzly Man*

Alf Seegert

I was in there the morning the Fish and Game officers were there examining the bear that had done the killing. The bear was all cut open. It was full of people. It was full of clothing. It was . . . We hauled away four garbage bags of people out of that bear. Treadwell was, I think, meaning well, trying to do things to help the resource of the bears. But to me he was acting like . . . like he was working with people wearing bear costumes out there instead of wild animals.

> —Sam Egli, helicopter pilot called to assist on the clean up after naturalist–film-maker Timothy Treadwell and Amie Huguenard were fatally attacked and devoured by a grizzly bear, *Grizzly Man* (Herzog)

And what haunts me, is that in all the faces of all the bears that Treadwell ever filmed, I discover no kinship, no understanding, no mercy. I see only the overwhelming indifference of nature. To me, there is no such thing as a secret world of the bears. And this blank stare speaks only of a half-bored interest in food.

> —Werner Herzog (Herzog)[1]

The clash of human obsession with the indifference of nature famously pervades director Werner Herzog's films. Recall Klaus Kinski's portrayal of the mad and manic would-be rubber baron in *Fitzcarraldo* (1982), who wants to build an opera house in the middle of the South American jungle and in order to do so conscripts hundreds of natives to drag a steamboat over a mountain, or Kinski's performance as the deranged conquistador whose Amazonian expedition dies off quietly one by one in search for El Dorado in *Aguirre: The Wrath of God* (1972). In these films, wild nature "revenges" itself on human beings simply by its implacability, its failure to fulfil the idealized fantasies of the obsessed.

Herzog's film *Grizzly Man* (2005) tells another story of obsession. Naturalist–film-maker Timothy Treadwell, after spending thirteen summers living with wild grizzly bears in Alaska—and recording over a hundred hours of video footage—is ultimately killed and eaten by a grizzly in 2003, along with his girlfriend, Amie Huguenard. Herzog's dour vision in *Grizzly Man* reinforces his conception of nature as starkly indifferent to human projects, but with a twist. As Herzog explains in his narration, "I discovered a film of human ecstasies and darkest inner turmoil. As if there was a desire in him to leave the confinements of his humanness and bond with the bears, Treadwell reached out, seeking a primordial encounter. But in doing so, he crossed an invisible borderline" (Herzog). As a passionate and sometimes manic seeker who risks his life to pursue his destiny in the wilderness, Treadwell cuts a distinctly Herzogian figure,[2] but unlike Kinski's darkly deranged figures in *Fitzcarraldo* or *Aguirre*, Treadwell more often resembles the naive and socially awkward protagonists of the films *The Enigma of Kaspar Hauser* (1974) or *Stroszek* (1976), both played by the real-life street musician Bruno S. Like these latter figures, Treadwell finds no place to belong within the complexities of human culture; he consequently cultivates an obsession not to overcome nature but rather to unite with it and leave human society behind.

But Treadwell's method of making contact with the creatures of Alaska's Katmai Peninsula departs radically from the standard tropes found in excursion narratives characteristic of environmental literature. Pastoral narratives typically promote the abandonment of technological interfaces in order for one to re-connect "authentically" with wild nature on its own terms: Witness Edward Abbey's Thoreauvian lament over the dissociating effects of electric generators, flashlights, and his "iron lung" of a trailer while living in the Arches in *Desert Solitaire*, or recall the "aesthetics of relinquishment" (Buell 143) in William Faulkner's *The Bear* in which Ike McCaslin sheds layer after layer of culture in order to become like the wilderness and make himself worthy to approach Old Ben the bear. In blinding contrast, Treadwell uses digital media to provide the very means of "making contact"—or at least for generating *perceived* proximity—with the creatures of the Alaskan wild. Treadwell's self-recorded video camera performances, often depicted in a visual field that he shares with the bears, are crucial to his project of making himself into kin (or at least virtual kin) of the grizzly bear. Likewise, Treadwell transgresses Herzog's "invisible borderline" precisely by means of the digital video camera and its footage—footage that represents nature and culture as if they inhabit the same world. *Grizzly Man* thus raises striking, eco-critically rich questions about the validity of technologically mediated representations that splice humans and nature together via visual framing, as well as about potential problems in the use of the technological inscription to "capture" nature in the frame in the first place.

To use N. Katherine Hayles's terminology, *Grizzly Man* is a "technotext"—a text, that is to say, that foregrounds the mode of its own making and the materiality of the medium from which it is made (Hayles 794)—in this case, videotape and film. The interface of the video camera functions not only as Timothy Treadwell's uncapped lens on the grizzly, but more importantly as a framing device for intimately linking himself with these bears. But the camera's very presence in wild nature also opens up possibilities for nature to resist such technological enframement, a resistance embodied in Treadwell's (and Amie Huguenard's) traumatic deaths-by-grizzly-bear. In an unthinkable rupture of the boundary separating the director and his subject matter, nature refuses to stay safely on the other side of the camera. That the camera moreover fails to fully represent their grisly deaths—the audio track records faithfully but the lens sees nothing, being capped—amplifies Herzog's view that nature exceeds the human capacity to represent it and belongs securely to its own world of sublime indifference to humanity.

Herzog and Treadwell thus embody competing visions of wild nature and its relation to human beings, as well as the role of technological mediation in representing this relationship (or, in Herzog's view, the lack of such a relationship). Treadwell uses the video camera as a medium of technological inscription and in so doing inscribes cinematic images that evoke not only his own subjectivity, but crucially, *intersubjectivity*: namely, an image depicting himself co-related and intimate with the bears, a relationship which Herzog considers in actuality to be no more than a one-sided, human projection. Herzog, in turn, remediates Treadwell's footage to make his own film, and does so in a way that rigorously reinforces his conviction that culture and nature are separated by an "invisible borderline" isolating them into two incommensurable, non-communicating domains. Is Treadwell's footage of himself among the bears a mere "special effect" that fictitiously splices together two radically dissociated worlds, as Herzog insists? Or might Treadwell's use of such a "machine in the garden" somehow become a form of mediation enabling actual "contact" with the more-than-human other? Overwhelmingly sceptical at such a possibility, Herzog focuses instead on Treadwell's gifts as a film-maker, acknowledging the astonishing natural beauty that Treadwell captures on film—but, notably, emphasizes this point when Treadwell is himself absent from the frame.[3]

Moreover, Herzog's use of Treadwell's over-the-top video-recorded theatrics ultimately transform his camp in the grizzly maze into a version of "camp" in Susan Sontag's sense, his presence enframed in quotation marks as taped performance pieces (Sontag). As a result, Treadwell's campy "wild" often recalls the artifice of Oscar Wilde more than it does the wildness of Alaska. Critics of *Grizzly Man* tend to debate questions over Treadwell's sanity, the legitimacy of his enterprise, and the ethical implications of grizzly bears becoming habituated to human presence (not to mention the fate of those bears which were fatally shot when recovery teams recovered Treadwell's and Huguenard's bodies). My interest here is not so much ethical as it is technological and aesthetic: What do Treadwell's and Herzog's rival conceptions of art reveal about human relationships with nature, and what do they suggest regarding our use of technology to frame wild nature in the first place? Does nature wield the final view, or do we? Validating Herzog's view that Treadwell bears the penalty of crossing an "invisible borderline," the breakdown of technological representation in Treadwell's death suggests that nature forever evades capture by technological inscription: Having been devoured by the very bears he once filmed, Treadwell becomes literally enframed by the wild creatures he himself once framed with the camera. Eco-trauma thus ensues by way of the bears' rival directorial and editorial control. With cold indifference, Herzog's nature thus revenges itself on Treadwell by means of its own merciless media of inscription, namely, teeth, claws, and paws.

NAMING AND FRAMING: MEDIATING KINSHIP THROUGH TREADWELL'S FOOTAGE

Early in *Grizzly Man*, Herzog presents viewers with a striking snippet from Timothy Treadwell's film footage: a grizzly bear cub approaches Treadwell's camera, and when Treadwell's hand reaches out to make contact, the bear sniffs at it and then—despite his voiced protests—persists in approaching him, threatening to unseat Treadwell and knock him over along with his camera. Shortly afterward, a subadult male challenges Treadwell and his camera, actually tipping them both backwards while the film runs. Instead of staying "on script," these bears not only break out of character as "stably framed objects" but also inadvertently demolish the "fourth wall" separating viewer from the viewed. In this way, Treadwell's footage yields paradoxically contrary effects, one suggesting connection with wild nature (Treadwell's outstretched hand sniffed by the inquisitive grizzly cub), the other dissociation (the grizzly's refusal to stop when asked). Were one of these bears a human actor, its behaviour might be construed as metatheatrical: a performer copping Calvino or miming Brecht and Pirandello, self-aware of his own performance and thereby exposing the evanescence of the line dividing the viewer from the viewed. But considered eco-critically, the young grizzlies' challenge to the technology of their filmic inscription suggests that wild creatures are more than passive objects of representation or "subjects of the gaze." Such an image for Herzog seems to be metonymic for nature's resistance to stable representation more generally, a resistance to being enframed in documentary footage.

But Treadwell's use of the camera is striking in how it unites through visual contiguity what (for Herzog, at least) proves an impassable gap ontologically. That is to say, it creates an image of humanity at one with wild nature. In another early scene in the film, two large grizzly bear males and Treadwell occupy the screen together, Treadwell in the foreground, grizzlies in the mid-distance snuffling in the grass. Treadwell speaks directly to the camera: "Behind me is Ed and Rowdy, members of an up-and-coming subadult gang. They're challenging everything, including me. Goes with the territory" (Herzog). This relatively seamless vision in which a human being shares the same frame with the massively more-than-human grizzly—no steel

bars between Treadwell's skin and the bears' fearsome "claws and paws"—
suggests an intimacy between them in a communally experienced landscape.
Contrast the typical Ansel Adams print in which humans are rigorously
abjected from the frame in order to let nature loom transcendent, uncon-
taminated by human society. In fact, the reference point for what counts
as "society" in Treadwell's films ultimately shifts from human society to
"human-grizzly" society, where both inhabit the same community.

Like his strategic visual framing, Treadwell's use of naming suggests intimate relations between himself and the bears, as we hear in the following
monologue.

> I'm here with one of my favorite bears. It's Mr. Chocolate. Hi, Mr. Choc-
> olate. He is the star of many people across the country: Children, people,
> adults. And we're here in the Grizzly Sanctuary. But I'm wrapping up
> my work in the Grizzly Sanctuary. Why is that? Because I'm on my way
> to the Grizzly Maze where bears do not have human protection, but are
> under human threat. Bears like Aunt Melissa. Bears like Demon, Hatchet,
> Downey and Tabitha. And it's time for me to go to protect them.
>
> (Herzog)

In these images, Treadwell not only appears in the same frame with the
bears (Figure 4.1), but also addresses them directly and by name (e.g., "Mr.
Chocolate"). Perhaps even more significantly, Treadwell invokes titles indicating family relations, for example, "aunt" and "Mr.," thereby suggesting
ties of kinship that link grizzly families to human ones. As Marsha and
Devin Orgeron argue in their essay "Familial Pursuits, Editorial Acts: Documentaries after the Age of Home Video," Treadwell is in effect shooting a
home movie for a family of his own invention (48).[4]

But are Treadwell and the bears in fact reciprocally *present* for each
other, or do they merely appear in juxtaposition with one another? Does his
footage qualify as a genuine "home movie" (i.e., with characters in his own
family) or instead as voyeuristic peepings by an outsider? A sceptic might
conclude that Treadwell's project here is, metaphorically speaking, merely
another version of those "pictures taken from the right angle" in which a
carefully positioned camera makes someone's hands appear to be holding up
the sun or where clever camera angles suggest obscene relations between a
body in the foreground and phallic-shaped geography or architecture in the
distance. Relations represented in such humorous snapshots are, of course,
purely rigged effects—at best, "virtual relationships."

Sam Egli, the helicopter pilot who flew in as part of a team to retrieve
Treadwell's and Huguenard's remains, echoes such a critique in asserting
that Treadwell wasn't really in contact with anything outside his own projections onto the landscape: "Treadwell was, I think, meaning well, trying
to do things to help the resource of the bears. But to me he was acting
like . . . like he was working with people wearing bear costumes out there

Figure 4.1 Activist and film-maker Timothy Treadwell stages scenes so that he appears alongside the bears he claims to protect in *Grizzly Man.*

instead of wild animals" (Herzog). Like the viewer of Herzog's *Fata Morgana* (1970)—a film that depicts both the beauty and the optical deceptiveness of mirages—Treadwell seems to delight in mistaking appearances for reality (though one might consider the possible deadly consequences if one construes a mirage for water in an actual desiccated wasteland and not merely on the screen). In *The Cinema of Werner Herzog: Aesthetic Ecstasy and Truth*, Brad Prager teases out problems with reconciling appearance and reality in German romanticism and examines how this conflict plays out in Herzog. For if we in fact see the world through "subjective lenses," then we are prevented from experiencing the world as it actually is (a point he raises in connection with the ruby glass in Herzog's *Heart of Glass*). Prager uses an example from Kleist: If our eyes were replaced with green glass, it would be true that the world appears green to us, but this would be because of our built-in mode of perception, not because of how the world is in itself (96). Likewise, it seems that the more Treadwell's vision is mediated by the camera lens, the more his experience of bears becomes a subjective perception dissociated from actual grizzlies-in-themselves. Treadwell's camera lens becomes a "green glass" that, through visual association, creates an appearance of intimacy with the bears not necessarily shared by those on the other side of the camera.

The tactic of naming, and Treadwell's domesticating language more generally (he says "go play" to the young bear who threatens to topple him and his camera), moreover anthropomorphizes the bears, which creates predictably mixed effects.

Anthropomorphizing the bears by definition makes them "more like us," which appeals enormously (for example) to the schoolchildren to whom Treadwell gives free presentations, and Treadwell is savvy enough as an activist to recognize the tactical value of such an approach (indeed, the name of the group he founded with Jewel Palovak is Grizzly People, a moniker which suggests both grizzlies as people and people as grizzlies). Moreover, most people only care for that which is "like them," as the charismatic megafauna that so regularly adorn Animal Planet episodes and World Wildlife Fund calendars dutifully attest. However, anthropomorphizing or "selfing" a bear's actions ("rub-a-dub-dubbing" is how Treadwell describes one bear's back scratching) thus risks denying the bears their irrevocable otherness as grizzly bears—rather than as "people wearing bear costumes." In his desire to become like the bears, Treadwell takes the tactic of "selfing" the grizzlies to a perilous extreme—namely, by giving them selves too much like his own. On the contrary, Herzog (as I shall argue) takes "othering" to an equally dangerous extremity in the opposite direction.

Perhaps most intriguing is how Treadwell not only names the bears and films himself in their presence, but in fact *performs* his alleged relatedness with these animals on film. When Ghost, the fox, runs off with Treadwell's hat in its mouth (presumably to lick salt from the headband?), Treadwell's taped response reveals in over-the-top stylings just how much he wants for this event to constitute not merely a raid by wild animal, but a prank by a naughty sibling:

> What are you doing to that hat? Where's that hat going? Hey, who's stealing that hat? Let me see that hat. Ghost, I want that hat. Man! Ghost is bad. Ghost, what are you doing with that hat? Ghost, that hat is a very important hat. Drop it! Hey! Oh, goddamn it! I can't believe this! Ghost! Ghost, where's that fucking hat? That hat is so frigging valuable for this trip. Ghost, you come back here with that friggin' hat. If it's in the den, I'm gonna fucking explode. Ghost, where's that hat? It's not okay for you to steal it. Oh, man! Oh, man! It's a friggin' den.
>
> (Herzog)

Treadwell here again achieves the appearance of kinship by using visual and audio imagery that splice the world of the fox and the world of Treadwell together.[5] Significantly, to all appearances Treadwell complains not merely *about* his hat's disappearance but rather *to* the fox who took it—another crucial demonstration of direct address that I want to take up in more detail later. Moreover, Ghost not only takes the hat but, in Treadwell's words, "steals" it, reinforcing the notion that Treadwell and fox share a commonly observed moral code—and that Treadwell can lay claim to reciprocal moral obligations from Ghost.

Treadwell's attribution of human-like motives to a non-human character recalls the child-like simplicity of Herzog's title character in *The Enigma of*

Kaspar Hauser, who enters society only after having spent all his formative years locked in a single room. When shown apples that have fallen to
the ground, Hauser (Bruno S.) comments on "how tired" they must be.
Hauser's benefactor attempts to correct him by insisting that apples have
no minds of their own and only obey physical laws when they fall—and he
attempts to prove it by rolling an apple on the ground, which another person will stop with his foot. The roll goes wild and the foot doesn't stop the
apple. "Smart apple!" says Hauser, confirmed in his animistic worldview.

One of the more peculiar methods of demonstrating kinship in Treadwell's
footage depicts him kneeling as he delicately cradles a pile of fresh bear scat.

> There's your poop. It just came out of her butt. I can feel it. I can feel
> the poop. It's warm. It just came from her butt. This was just inside of
> her. My girl. I'm touching it. It's her poop. It's Wendy's poop. I know
> it may seem weird that I touched her poop, but it was inside of her. It's
> what . . . It's her life! It's her! And she's so precious to me. She gave
> me Downey. Downey's . . . I adore Downey. Everything about them is
> perfect.
>
> (Herzog)

What might seem like an unprecedentedly bizarre eco- scat-fetish in fact
has considerable literary and critical precedent. Treadwell's actions here
reprise those of young Jean Jacques Rousseau, as he seeks to experience the
unmediated presence of Maman, his beloved foster mother, which Derrida
discusses in *Of Grammatology* (152). Rousseau kisses the floor and the
coverlets that she has touched, and at one point even resorts to telling her
that she has contaminated food in her mouth so that when she spits it out
he can then consume it himself. Rousseau (mistakenly) thinks that if only he
can get close enough to Maman, he will at last overcome his every sense of
lack and finally bask in her full presence. Treadwell similarly wants perfect
union with the bears—and to that end, what was inside Wendy the bear he
touches adoringly as a surrogate trace of her untouchable interior.

But as with Rousseau and Maman, the recorded *performance* of such
desire is especially crucial because it authenticates Treadwell's intimate connection with Wendy and the other bears in the first place. By registering
such seemingly bizarre engagement with bear scat on film (bizarre from a
human standpoint, that is), Treadwell is performing a tellingly bear-like act.
Acting himself like another bear, Treadwell treats scat as a meaningful *sign*
demanding a visceral interpretive response—and not merely as something
to avoid stepping in. Scat and urine, in the world of the bears, are identity markers telling secrets about their interior worlds. (In Stefan Quinth's
short 2005 documentary *Deadly Passion: Tragedy in Katmai*, Treadwell is
described as even smearing bear dung all over himself so as to become more
bear-like.) Treadwell's palpable *yearning* in this scene (which, as my student
Alan Reiser points out, seems overwhelmingly charged with erotic desire)

moreover suggests that he himself wants not only to connect with the bears, but actually to consummate his relations with them as another bear himself.[6]

Treadwell's use of language in the scene with "Wendy's poop" is particularly telling in its suggesting Treadwell's familial—perhaps even fatherly—role. But has Wendy in fact "given" him Downy? Is Treadwell in any way the object of the bears' awareness at all? One is tempted to wonder if Treadwell in fact considers himself the patriarch of this "family,"[7] but his ebullient and earnest performances on camera suggest that rather than merely documenting his relationship with the bears, the camera's footage in fact plays a role in *creating* such a relationship—much like Judith Butler's notion of gender as something which comes into being by way of repeated performance rather than as a pre-existing essence (Butler, *Gender Trouble*). Treadwell's camerawork thus tweaks Derrida's logic of supplementarity in a highly intriguing manner. Denied "withness"—for he can never be fully "with" Wendy as part of her ursine family unit—Treadwell instead uses the camera as *witness* to create the semblance of family on film. In this respect, the camera even assumes the role of a divine, vindicating omniscience that recognizes his true status as a member of the grizzly community. Addressing the camera, Treadwell says "if there's a God, God would be very, very . . . pleased with me. If he could just watch me here, how much I love them, how much I adore them, how respectful I am to them. How I am one of them" (Herzog). As Marsha and Devin Orgeron put it, Treadwell "seems to lose track of his existence outside of the camera's presence, needing it as a witness to these intimate moments, even the flawed moments with his unconventional, interspecies family" (58). In a similar vein, communication scholar John Durham Peters reminds us moreover that to be a witness to something requires not only that one see, but that one speak about what one has seen to another (Peters, "Kittler Seminar"). The camera fulfils this role by acting as more than mere video recorder. Because of the easy reproducibility of the cinematic image, the camera provides for ready dissemination of all that it views. Scenes like the one where Treadwell apostrophizes to the camera over Wendy's poop thus foreground the importance of the video camera as Treadwell's confessional—one who sees and validates Treadwell's actions as a communicative medium that ultimately tells all.

The camera eye, as mediating lens, thus arguably has as much of a constructive role as it might have of merely recording or documenting nature itself. Along these lines, N. Katherine Hayles insists that through their pervasiveness and the textual inscriptions they produce, technologies such as the video camera ultimately *constitute* the subjects they film. In "Remediation in *House of Leaves*," for example, Hayles argues that Mark Danielewski's novel

> instantiates the crisis characteristic of postmodernism, in which representation is short-circuited by the realization that there is no reality independent of mediation. Rather than trying to penetrate cultural

constructions to reach an original object of inquiry, *House of Leaves* uses the very multilayered inscriptions that create it as a physical arti-fact to imagine the subject as a palimpsest, emerging not behind but through the inscriptions that bring the book into being.

(Hayles 779)

If the characters from *House of Leaves* possess *subjectivities* that emerge only via Hi8 video and Zampano's textual collages (these being themselves further remediated by the narrator Johnny Truant), then I am led to ask if something partly analogous but crucially distinctive might be happen-ing through technological mediation in *Grizzly Man*. In *Grizzly Man*, Treadwell's subjectivity emerges through images that are primarily meant to evoke *intersubjectivity*—or at least a simulacrum of intersubjectivity. Under this logic, grizzly bears and Treadwell share footage, and therefore share a world. Treadwell thus forges a family not through blood but through media.[8] Echoing the boundary-manglings enacted by Donna Haraway's cyborg (Haraway, *Simians, Cyborgs, and Women*), Treadwell hybridizes a family made up of humans, grizzlies, and—crucially—digital video technol-ogy. For Treadwell the camera "watches over" him and the grizzlies like one of Richard Brautigan's "Machines of Loving Grace" (1968). The last stanza of that poem reads

> I like to think
> (it has to be!)
> of a cybernetic ecology
> where we are free of our labors
> and joined back to nature,
> returned to our mammal
> brothers and sisters,
> and all watched over
> by machines of loving grace.
> (Brautigan)

Brautigan's poem is ambiguous on the question of whether our being "joined back to nature" as "brothers and sisters" actually *requires* that we be "watched over" by such machines. For Treadwell, however, the machine performs the very splicing needed to heal the rift dividing a more-than-human set of family relations. In so doing, it moreover witnesses and vali-dates Treadwell in his unique, self-appointed role as the protector who must in turn watch over the grizzlies himself. For Herzog, however, Treadwell's project of uniting humans with nature via a shared world of footage is like trying to heal a fractured bone by photo shopping an X-ray image of it. What happens in art does not magically reproduce itself in life. The technol-ogy of the camera, as Herzog quickly demonstrates, can as easily dissolve bonds as forge them.

REMEDIATING TREADWELL: HERZOG'S
KINSHIP-DISSOLVING LENS

In *Grizzly Man*, nature is not only *mediated* variously through Treadwell's footage, through Herzog's footage, and through the vantage point of several interviewees, but also (in Bolter and Grusin's terminology) *remediated* (Bolter and Grusin). This is to say that the bears are first rendered into Treadwell's hand-held footage, and this footage is in turn appropriated and re-presented by Herzog for his own film. Although many critics and viewers refer to *Grizzly Man* as a documentary, Herzog disparages the terms "cinema vérité" and "documentary" alike, citing that these offer merely the "accountants' truth"—that is, mere correctness or factuality—rather than the "ecstatic truth" of art, so he makes no sharp distinction between his "feature films" and his "documentaries." In the essay "Conceiving *Grizzly Man* Through the 'Powers of the False,'" Eric Dewberry points out that "as a film-maker, Herzog shares this ideological perspective: all documentary is false even if it conveys the myth of objectivity. The function of the filmic image for Herzog is not to represent reality, but rather to build and shape images to form a facet of unseeable and unsayable truth" (2). As Herzog himself puts it in his "Minnesota Declaration," a manifesto against the "authenticity" of so-called documentary filmmaking, "There are deeper strata of truth in cinema, and there is such a thing as poetic, ecstatic truth. It is mysterious and elusive, and can be reached only through fabrication and imagination and stylization" (Herzog, "Minnesota Declaration" n.p.).

Herzog's "documentaries" are thus filled with imaginative stylizations and often outright fabrications, and his so-called feature films often use real props on authentic locations rather than simulated effects, further blurring the line between the two. (For instance, the ship towed over the mountain and later dashed on the rocks in *Fitzcarraldo* was a real steamboat, not a model.) Herzog not only directs *Grizzly Man*, but at times also acts as a literal mediator between image and audience, not only narrating but even inserting himself bodily into the film. Because the filmic medium is fully under his directorial control, Herzog makes viewers fill in whatever information he chooses to eclipse with his own presence—and the audience soon discovers that even small trickles of bandwidth will be administered through the intercession of the film's director-priest. Perhaps the most striking moment in all of *Grizzly Man* occurs when Herzog simultaneously reveals and conceals the audio recording of Treadwell's and Huguenard's fatal mauling. Instead of directly playing back the tape for the audience to hear, Herzog (wearing headphones) relays to Jewel Palovak, Treadwell's friend and exclusive keeper of the tape, what he experiences in listening to Treadwell's and Huguenard's final moments. The audience witnesses Herzog's remediation of the footage in real time. Herzog begins by relating content such as "I hear rain, and I hear Amie, 'Get away! Get away! Go away!'" but before long, when his hand wearily reaches up to cover his face, his physical reactions suggest more than his words do. He asks Jewel to turn off the tape, insisting, "Jewel, you must never listen to this [. . .] I think you, you should not keep it [. . .] it will be the white elephant in your room all your life" (Herzog). Ultimately, he instructs Jewel to destroy it.

In this scene the video camera's very *failure* to represent—caused by the lens cap's obscuration of the video while the audio still records— intriguingly becomes the vehicle for amplified storytelling. Herzog's yearning for "ecstatic truth" over "the accountant's truth" is fulfilled by the very inability of the visual image here to denote anything. One might construe Herzog's mediating obscurations as compensating for the cloudless serenity of the Alaskan wilderness, which might otherwise be at risk of seeming too transparently pretty (or its creatures, too "cute"). Brad Prager argues a similar point in relation to the visual style of Herzog's 1976 film *Heart of Glass*: "Its concealment of its landscape beneath a veil of fog and twilight [. . .] resist[s] the cheery mood and bright colours associated with Heimat films" (94). (Heimat films promoted images of unspoiled nature as "homeland" in postwar German cinema.) Herzog's remediation of nature via headphones not only adds mystery, but also deflates the possibility for nature to be sentimentally domesticated via the visual image. It might not be too much to say that Herzog perceives Treadwell's films as, in effect, Heimat versions of nature that his own film needs to rectify. Instead of being spoon-fed cinematic images of sublime arctic scenery, Herzog's viewer must here work to imaginatively fill in the gaps, reader-response style, to overcome the impasse of halting representation. But it is not enough that the viewer do this work—instead, Herzog must himself mediate the delivery of the audio track, providing his own cues and responses to ensure that the desired results obscure all capacities for denotatively secure representation.[9] Ironically, the sensory limits of the headphones minus visuals allow Herzog to transcend representation and thereby engineer an experience of the sublime.[10]

Although the audio track of Treadwell and Huguenard's fatal mauling is in principle indefinitely replicable, Herzog demonstrates a palpable need for this recording never to be reproduced. Why does Herzog instruct Jewel to destroy it? Because it is too horrible? Or perhaps does he do so to maintain the aura of the tape (and his privileged access to it) by destroying all potential copies along with it? The camera's disabled visual sensations, combined with Herzog's verbal remediation of the audio track, themselves generate an aura of mystery, enabling the footage to escape from the fate Walter Benjamin ascribes to mechanically produced media in which aura fades with the loss of a unique "original." By thus remediating the audio track into halting signifiers of his own production, Herzog achieves the same hypnotic effects he obtained by making his camera linger over misty landscapes and waterfalls in *Heart of Glass* (1976)—namely, the evocation of an eerie, cloud-shrouded terrain of sublime and hidden depths. In that same film, Herzog famously hypnotized his cast during filming to bring forth a more "authentic" performance, and at one point even contemplated introducing the film with a self-proclaimed attempt to hypnotize his viewers (Prager 96).

"Ecstatic truth" demands some measure of transcendence of conditioned, rational responses, and hypnosis is one way to achieve such an escape. But whereas Treadwell fell under the irrational spell of the grizzlies, we as an audience are meant to swoon before the artistic spell of the director rather than succumb to Treadwell's call of the wild.

Though one might be tempted to read Herzog's refusal to "cross the line" and share the audio track with the audience as a gesture of deference to that audience, such calculative obscurations span Herzog's career and suggest that something more significant is going on than merely sensitivity to Treadwell, Huguenard, Palovak, or his viewers. Other instances of similar withholding of information in Herzog's films include the reporting—but not displaying—of the "sacred object" in *Where the Green Ants Dream* (1984) and the signature depiction of fog-covered landscapes in *Aguirre: The Wrath of God*, *Heart of Glass*, and more recent works like *Rescue Dawn* (2007). Herzog's "documentary" film *The White Diamond* (2004) offers perhaps the most striking analog to the headphone-remediation scene in *Grizzly Man*. In the film, Herzog follows aviation engineer Graham Dorrington, who wants to fly a specially designed airship over the rainforest canopy in Guyana, South America. The launch site is near the massive Kaieteur Falls. Below this vast curtain of water up to a million swifts gather and flit about each day eating insects, and behind this waterfall all the swifts gather each night to roost, invisibly. The site is inaccessible to people and, as a result, has become a focal point of local legend. Curious, Herzog sends the team's physician/mountain climber over the cliff's edge to see if he can film the secret behind the falls. He does so, but Herzog tells the audience at the end of the film that he will not show the footage and instead only shows the camera spinning in circles as it is pulled back up to the rim of the cliff. The swirling image leaves the audience vertiginous. At the end of the movie, Herzog dedicates the film to this "secret kingdom of the swifts," and his withholding of information ensures that their domain will remain a tantalizing secret to his audience. Herzogian eco-trauma in *Grizzly Man* thus works according to the logic of trauma outlined by Joshua Hirsch in *Afterimage: Film, Trauma, and the Holocaust*. Herzog, like posttraumatic cinema more generally, formally mimics the post-traumatic experience itself in an attempt "to formally reproduce for the spectator an experience of suddenly seeing the unthinkable" (19). The caveat, of course, is that in *Grizzly Man* this traumatic image is never made present except through the imagination, which renders it that much more horrifying.

Herzog's remediation of Treadwell's and Huguenard's final moments evokes not only aura and mystery, but also a chilling image of nature that inverts the standard mode of enframement imposed on nature by cinematic media. The image is ostensibly something *produced*, whereas the object of representation is *consumed*. But in *Grizzly Man*, nature chews up this model and violently spits it back out: The represented (nature) instead consumes the representer (Treadwell) in an action itself only haltingly representable (no video, and only Herzog-mediated audio). The blinded camera and Herzog's remediation of the audio track together insinuate that this "primal encounter" (Herzog)—namely, nature's inscribing and consuming of Treadwell and Huguenard—cannot be adequately captured through media. Rather, nature is the one that captures and consumes the image-maker. Herzog thus secures the inscrutability of nature and its utter separation from the human symbolic order. To be sure, for Herzog nature can still be represented through cinematic art, but in his view such representations should aim at "ecstatic truth" rather than seek to document facts. Such images are misleading if read literally, or in Herzog's own terms, when taken as "the accountant's truth." For example, in the DVD extras for *Grizzly Man* one sees Herzog instruct the musicians as to the appropriate style of music for what might appear an idyllic and playful scene of Treadwell and a grizzly bear swimming together. Herzog insists that the appearance of unity with nature in this scene is deceptive, and that the music should rather convey the irreconcilability of human and wild worlds despite all appearances to the contrary. The tones, harmonically ambiguous, never suggest a major key.

Treadwell's use of his own video footage to demonstrate his intimacy with the bears not only raises such issues of "ecstatic" versus the "accountant's" truth, but also calls out for a reading in relation to a long-standing tradition of pastoral art and literature. In *The Machine in the Garden*, a well-known study of technology and the pastoral ideal in America, Leo Marx argues that pastoral narratives traditionally attempt to occupy the "middle landscape" between urban and wilderness worlds, halfway between culture on the one side and savagery on the other (139). With this conception in mind, one might conceivably read Herzog's "invisible borderline" as demarcating the liminal zone inhabited by Treadwell on camera, to his own hazard: a pastoral enclosure reconfigured as the bounded cinematic frame harmoniously inhabited by both Treadwell and the bears. Inside the frame, pastoral artifice dominates in the form of Treadwell's directorial stylizations, his self-conscious performance as a character, and his arrangement of the "players" to make sure that human and bear occupy the field at the same time.

Of course, not all versions of pastoral are naïvely ideal or romanticized: Terry Gifford reminds us that strains of the antipastoral have haunted the pastoral at least as far back as the poets John Clare and George Crabbe (116)—if not Virgil—and have led to a more nuanced "postpastoral" (146) in more recent writers like Edward Abbey, whose Eden contains rattlesnakes, quicksand, and the bloated corpse of a sightseer who loses his way. Treadwell himself repeatedly recites, almost like a mantra, that these bears "can kill, they can decapitate," so he seems more than just nominally aware of the dangers he faces as well. Treadwell's doom thus comes about not merely by his underestimating the voraciousness of his cinematic characters, but rather by misconstruing the narrative landscape—or lack thereof—which together they occupied. For even the most nuanced pastoral narrative still subscribes to an *aesthetic* mode of representation, which is to say that it attempts to reconfigure nature in terms amenable to artful storytelling. But for Herzog, although nature can be represented aesthetically, nature is not itself an aesthetic mode of being.

To appropriate the narratology of Russian formalist Viktor Shklovsky, one might say that Herzog's nature is at best a version of *fabula*, or mere chronological narrative (Shakespeare's "tale told by an idiot, filled with sound and fury, signifying nothing"), a narrative in which one thing follows another in brutish succession, all governed by the blind necessities of hunger and reproduction, not by artfully arranged connections leading in any meaningful aesthetic direction. Herzog's nature equates to, in his own words, "harsh reality" and "overwhelming indifference"—and ultimately, death (Herzog). Treadwell, as both character and director, in contrast frames the wild in terms of *syuzhet*, or plot. After a long soliloquy in which he thanks the bears for giving his life purpose—which itself suggests the motivations of plot and character in nature's supposedly having "given" something to him— he abruptly switches modes like a newscaster turning to a reporter in the field: "Now let the expedition continue. It's off to Timmy, the fox.[11] We've gotta find Banjo. He's missing!" (Herzog). Treadwell's storied landscape thus has distinct protagonists (here, Banjo and Timmy) as well as antagonists (including rogue bears, the drought, or the wolves who prey on Treadwell's fox-friends), which suggests that the motives of interested characters run the show and not merely blind instinct.[12] Likewise, when Treadwell witnesses a vicious fight between two male bears competing for a mating partner, he creatively reinterprets the aftermath by invoking the cadences of Howard Cosell commenting theatrically on the outcome of a prolonged boxing match:

> Here I am at the scene of the fight. It looks as if tractors tore the land up, raked it, rototilled it, tossed it about. There is fur everywhere, and in the camera foreground excreted waste. In the middle of the fight so violent, so upsetting that Sergeant Brown went to the bathroom, did a number two during his fight. Extremely emotional, extremely powerful.
>
> (Herzog)

Treadwell's language quickly shifts to chivalric, however, where he re-conceives the fight as a battle for the honour to "court" Saturn, whom he dubs "the queen of the Grizzly Sanctuary." (Considering Treadwell's "Prince Valiant" haircut and his former job as a "squire" working with Jewel Palovak at the theatrically themed restaurant Gulliver's, it is not so surprising to witness him projecting such stately virtual overlays onto nature.) Most fascinating, however, is when Treadwell, again like a sports journalist, actually *addresses* Mickey the bear, who lies on the ground, seemingly exhausted.

> I just wanna discuss that fight with Mickey bear right here. He's right next to me here in the Grizzly Sanctuary on the tide fly. Saturn off to camera left. Mick, you underestimated Sergeant Brown. You went in for the head. He seemed to be rope-a-doping you like he wasn't that tough.

And then once you banged into him, man, he turned out to be one heck of a rough bear, a very rough bear. I was so scared, I almost got sick to my stomach watching you fight. Then when he knocked you down and you were down on your back, it was terrible, it was terrible! I'm not duking it out for any girl like that.

(Herzog)

Treadwell concludes by imaginatively identifying with Mickey in desiring Saturn. "Well, I've had my troubles with the girls. Yeah, yeah. And I'll tell you something. If Saturn was a female human . . . I can just see how beautiful she is as a bear. I've always called her the Michelle Pfeiffer of bears out here" (Herzog). Treadwell's reading of the landscape and his "dialogue" with its creatures thus proves endlessly creative in its artful fabrications.[13] Herzog himself approves of such fabrications that challenge denotative truth and serve the ecstasy of art. But if I read Herzog's critique correctly, what Treadwell failed to recognize is that although art can freely transform nature into whatever ecstatic truth he might imaginatively forge, nature itself is not—and can never be—art. The Katmai Peninsula and its wild creatures ultimately belong to themselves, not to an artfully framed "scopic regime." Treadwell's transgression was thus to project art onto the wild and then cross the line dividing them in order to inhabit the virtual wilderness of his imagination, Quixote-like, in the real world. But whereas Cervantes' virtual giants overlaid harmless actual windmills, Treadwell's imagined kindred are hungry and come equipped with built-in claws and paws. Pastoral, as the idealized middle space between wilderness and culture, is thus sustainable, but only as an image, a virtual space on film. When Treadwell attempts to actualize such ideals by living in the kind of intimate harmony that his films suggest, the genre flips channels from naïve pastoral (which includes bears with names like "Thumper") to nameless horror (literally "nameless," for there is no indication that the bear that attacked and ate Treadwell was one he even recognized). Disney, meet Lovecraft. As Herzog says in his 1999 film *My Best Fiend*, if nature is indeed a harmony, then it is "the harmony of overwhelming and collective murder." Six years later, in *Grizzly Man*, his view has not softened: "I believe the common denominator of the universe is not harmony, but chaos, hostility, and murder." But Herzog's coherence here is questionable. Even as he criticizes Treadwell for projecting human traits onto the wild landscape, Herzog himself freely employs terms like "murder" and "fornication" (Herzog)—culturally framed ethical assessments—when accounting for natural phenomena like predation and mating, respectively.

AESTHETICS VERSUS COMMUNICATION: HERZOG, TREADWELL, AND THE FUNCTION OF CINEMATIC ART

Although one might conclude that, based on the preceding comments, Herzog and Treadwell disagree primarily in their views of nature, even more decisive are their diverging conceptions of art. In many respects, Treadwell appears to be an artist only by accident. In one of his most striking and evocative images, the mounted camera lingers to witness the unhurried blowing of willows for fifteen seconds, Treadwell having absented himself from the frame. Herzog comments, "In his action movie mode, Treadwell probably did not realize that seemingly empty moments had a strange, secret beauty. Sometimes images themselves developed their own life, their own mysterious stardom" (Herzog). Such scenes, in which nature moves to its own rhythms and under its own sway, cast a spell over the viewer. Treadwell's shots here rival those of Herzog's own camera, which lingered over the churning Urubamba River in *Aguirre: The Wrath of God* and which levelled its gaze at the tumbling waterfall in *Heart of Glass*. Treadwell's images of vegetation tossed by the wind also echo, probably inadvertently, the sunset-lit swaying of Terrence Malick's grasses in films like *Days of Heaven* (1978). But at the very moment the viewer is tempted to succumb to such hypnotic lulling, Treadwell himself jumps into the frame, transforming the cinematic idyll into action romp, saying "Starsky and Hutch. Over" (Herzog). The filmic reverie only happened, we discover, because of pragmatic necessity: Treadwell had to leave the camera rolling on the tripod while he left the scene in order to dramatically re-enter it. The Malick-evoking vegetation-in-sway would surely never make it into Treadwell's own final cut, however much Herzog foregrounds it in his own.

Treadwell and Herzog thus diverge with respect to the very *uses* to which they put nature, artistically speaking. For Herzog, however beautiful it might be, life in nature still equates to hunger, mutual predation, and death—a harshness only tolerable if nature remains utterly impersonal and other; hence, "we ought to be grateful that the Universe out there knows no smile" (Herzog, "Minnesota Declaration" n.p.). If nature in fact had a smile, one imagines that for Herzog it would grotesquely mask malicious intent, like the smirking face of Heath Ledger's Joker. But despite its horror, nature holds power for Herzog as a repository for images that, in an able film-maker's hands, yield aesthetic and ecstatic truth. But for Treadwell, nature does not exist primarily for the sake of art. If Herzog's nature is *setting* that demands cinematic representation by a director, for Treadwell nature is *character* with whom we are called to seek dialogue as fellow characters in a shared story.[14] Whereas Herzog seeks to fulfil the demands of *aesthetics*, Treadwell aims to fulfil the demands of *communication*. This point is crucial because, as Terry Eagleton points out, literary art inheres precisely in using words to do more than just communicate—and the same is true with the use of images in cinematic art. Thus, for Herzog, Treadwell's desire to constrain his images to the service of real-life interaction risks exorcizing ecstatic truth in favour of terms amenable to the accountant.

Treadwell's footage depicting himself interacting with the bears raises a challenge not only to Herzog, however, but also to wildlife biologists and makers of nature documentaries, who typically seek to obscure their own presence in the landscapes they film, thereby conveying an illusion of immediacy and transparency that represents "pure nature" free from human interference. But Treadwell has discarded his objective lens in favour of a subjective one. Treadwell's greatest "boundary violation"—if violation it is—inheres in the fact that he not only represents himself *in* nature but actually talks *to* nature. In so doing, Treadwell (consciously or not) in fact reprises the work of earlier visionaries. In *In the Remington Moment*, Stephen Tatum points out how the trope of "the call" in the early twentieth-century artwork of Frederic Remington demonstrates an urgent attempt to "register this particular artist's and his dominant culture's deep-seated anxiety about losing contact with some version of authentic, real life located elsewhere than in urban industrial society" (61). Like Herzog, Treadwell seeks an ecstatic truth—but for him this truth can exist in literal conversation, not just in art. When Treadwell chides Ghost the fox, addresses Mr. Chocolate, interviews Mickey, or reprimands the unruly bear cub on film, he engages in a mode of representation that transcends existing conventions that reduce nature to mere *object* of representation. Philosopher David Abram argues that this species of engagement is far from delusional, and in fact embodies yet another form of ecstatic truth, namely, *prayer*.

> In its oldest form, prayer consists simply in speaking *to* the world, rather than solely *about* the world. We should recognize that it is lousy etiquette to speak only *about* the other animals, only *about* the forest and the black bears and the storms, since by doing so we treat such entities as totalize able objects, able to be comprehended and represented by us, rather than as enigmatic powers with whom our lives are entwined and to whom we are beholden. Can we not also speak *to* these powers, and listen for their replies?
>
> (Abram 189)

In response to Abram's question, Herzog would say "no" because for him, nature's sounds are inarticulate and signify nothing: "Mother Nature doesn't call, doesn't speak to you, although a glacier eventually farts. And don't you listen to the Song of Life" (Herzog, "Minnesota Declaration" n.p.). To speak to nature and listen for a reply would thus be, in Herzog's view, a category mistake. Such a thing might be possible through art—recall the charming animism of Kaspar Hauser—but it exists only in the ontology of imagination, bounded firmly by an aesthetic frame.[15]

But is Herzog's own elevated conception of art and (arguably) reductive view of nature any less philosophically problematic than Treadwell's naïve visual realism? For all their differences, both directors use video technology to frame the bears for particular ends, and in so doing literalize Martin Heidegger's notion of technology as "enframing" (*Gestell*). For Heidegger, technology is itself "nothing technological" but foremost a mode of revealing (287). Unlike other modes of "bringing forth" (*poiesis*), however, technology "challenges forth" nature into particular modes of production readily consumed by human beings, and in so doing, threatens to reduce Being to *Bestand*, or standing reserve: earth-as-quarry awaiting excavation in order to serve exclusively human ends. Perhaps more significantly, technology, aided by what Heidegger laments as the modern impulse to conceive the world "as picture," conflates Being itself with the human representations that "frame" it in terms of its human use (think of the use-value couched in such terms as "natural resources," "scenery," and even "lodgepole pine," for example). In consequence, we readily identify nature, for example, with our techno-cultural modes of *representing* nature in the first place, finding in nature what we ourselves put there.

Such a predicament forces us to ask questions about our use of technology to produce images. For example, Kevin DeLuca, author of *Image Politics*, asks us to consider the orienting question "What sort of relationships to the earth and world does a technology enable?" (78). More specifically, we might ask what the consequences are of a scopically oriented technology such as the video camera on our experience of nature. Is Treadwell's footage an enframing that "challenges forth" a representation of the grizzly that reduces it to an image "awaiting human consumption?" And if so, what sort of consumption is being served? Does video technology inevitably lead to an image of nature as something that exists only "to be consumed?" Image-producing machinery might seem out of place in the pastoral garden, but as DeLuca points out, its use is far from alien in the service of environmental activism: "One rhetorical strategy that all stripes of environmentalism have embraced" is that of "deploying images to save nature" (81).

One way to approach the question concerning technology and Treadwell might be to ask if Treadwell in fact "challenges-forth" a particular representation of the grizzlies, or if he might be said rather to "dwell" with them. For Heidegger, "dwelling" implies that one lives in such a way that the essence of a being is revealed in its own fullness rather than being re-channelled in service of reductive, or specifically human, ends. The word "challenge" is problematic in this case, however, because Treadwell sees the act of challenging as in fact *characteristic* of the grizzlies themselves: As he says about the subadult males who challenge everything at the beginning of the film, including him, it "goes with the territory" (Herzog). Ultimately, even his own death-by-grizzly does not undermine Treadwell's claim to have been one of them himself; grizzlies kill and eat one another routinely. Nick Jans thus considers Treadwell's bear-mediated demise as perhaps an "ironic compliment to a man who strove, among bears, to become as much like them as possible" (215). Treadwell claimed to use his grizzly films to educate school children and to raise social awareness about threats to grizzlies from poaching and habitat loss. But, as Herzog informs his viewers, most of the bears that Treadwell rose up to "protect" were already living in federally protected areas. Treadwell's footage thus seems to serve a far more personal agenda than that of raising general ecological awareness. As we have seen, Treadwell uses the images in his films to create the sense that he is himself part of what he represents, a member of the grizzly family. In a perverse irony, however, Treadwell's use of exclusively human technology becomes Treadwell's very means for *escaping* his humanity. Through his artful deployment of the video camera, Treadwell fuses art and reality together to make something simultaneously beautiful and monstrously inhuman. As a result, Treadwell and the bears now dwell together forever in a digital ecology as mass-produced images etched onto the infinitely repeatable play surface of Herzog's DVD, haunting a fictive peaceable kingdom in the light of an eternal summer. Just as all of the world's dead wait for the needle to drop yet again onto Nietzsche's eternal gramophone record in a grand "eternal return," Treadwell and his bears await their digital re-summoning by a spinning of a disc and the caress of a laser beam.

But must Treadwell's imagined grizzlies really only wander the spectral landscape of his fancy? Does he in fact cleave to a clan of "virtual grizzlies" like he clings to the childhood teddy bear that shares his tent in the Grizzly Maze (Figure 4.2)?[16]

As Herzog says in the film, the bears do more to save Treadwell than he does to save them. But it's not only Treadwell's lens that does the framing in *Grizzly Man*. As DeLuca points out, different uses of technology are hooked into different technological regimes—and Herzog's own use of film and camera reveal a particular agenda in what they "bring forth." When Herzog says that in the bears he sees "no kinship, no understanding, no

Figure 4.2 Timothy Treadwell pictured with his stuffed teddy bear in his tent during a rainstorm.

mercy" but only "the overwhelming indifference of nature" (Herzog), is he not projecting and framing as well? To say, as Herzog does, that "there is no such thing as a secret world of the bears" (Herzog) denies the bears any point of view of their own, and rules out the possibility that they had their own inscrutable regard for Treadwell during those thirteen summers in which he lived so closely with them. DeLuca reminds the viewer that not only Herzog's ecstatic truth matters, but scientific truth as well. The bears, which have poor vision but keen olfactory senses, would have in fact been acutely aware of Treadwell's smell for thirteen summers and in that respect they did in fact share worlds (in a remarkably tolerant way, it might be noted). Because of this fact, DeLuca argues,

> Treadwell's smell makes him very much a part of their world. There is very much a relationship between Treadwell and the bears (what kind is open for debate). They know he is there and react to his presence. It is not a relationship fabricated by camera frames and angles.
>
> (DeLuca, letter to the author)

But to ensure the sanctity of aesthetic truth, Herzog in effect Orientalizes the natural world, securing its otherness at the cost of denying it a corresponding selfhood of its own. He thus reappropriates Karl Marx's dictum on the so-called Orient, applying it to nature and wild creatures: "They cannot represent themselves; they must be represented" (qtd. in Said xiii).

As director of *Grizzly Man*, Herzog likewise wields the enclosing frame that contains Treadwell himself. In Herzog's hands, the question of Treadwell becomes ultimately a question about which encloses what: Is art the bigger frame, or nature? Like Oscar Wilde's Dorian Gray, who trades his image with one on the canvas, Treadwell occupies the cinematic frame as a living performance piece, throwing himself, as it were, "into the Wilde." But like the protagonist of Wilde's novel, Treadwell's character tends deathward in direct proportion to his exceeding the boundaries of his artistic enclosure. When Dorian Gray ultimately holds his own artistic project accountable to the claims of morality, he is slain and takes on all the disfigurements which had heretofore been displaced to the canvas. Treadwell, too, survives when to all appearances he shouldn't, seemingly immortal, so long as he dwells exclusively in the space of art, which for him consists in the pastoral enclosure of the cinematic frame. But like Dorian Gray, Treadwell finds only death when he holds cinematic aesthetics accountable to the depredations of nature lurking just outside the boundaries of art. In turn, his audience shares in this version of eco-trauma. The enclosing body of the bear replaces the quotation marks of Treadwell's full-time artistic performance, suggesting that nature and death hold the final account, together measuring the ultimate frame that encloses the world—the very frame Herzog himself seeks to enclose through the aesthetic ecstasy of film-making.

NOTES

1. All references to "Herzog" with no specified title are to his film *Grizzly Man*.
2. In "Grizzly Tale: Werner Herzog Chronicles Life of Ill-Fated Naturalist," Maria Garcia ponders Treadwell's relationship to Kinski's characters: "Timothy Treadwell may be the newest addition to Herzog's oeuvre, but in his eccentricities, borne of a troubled fife, he bears a striking resemblance to the characters portrayed by Kinski in Herzog's narrative films. Asked about this cinematic concatenation that reaches back to Kinski as the mad conquistador Aguirre or as the iconoclast Fitzcarraldo, Herzog chuckles. 'I think the similarity between characters like Kinski in *Aguirre: The Wrath of God* and Treadwell in *Grizzly Man* is there,' he admits. 'Yes, it's true, of course. It is as if they were in the same family, as if I was working on a big family series, like doing 'Bonanza.' But I am not searching for these characters. They just come across me'" (Garcia n.p.).
3. Though Herzog roundly criticizes Treadwell for his numerous boundary violations, Herzog's view should not be equated with dismissal. His disagreement is, in his own words, "not a violent argument," but rather "the same way I argue with my brothers." Extending his praise for Treadwell's craft, Herzog continues, "I have to give Timothy credit. He created footage that is unprecedented in its beauty. I salute him as a wonderful fellow filmmaker." Regarding his own role in directing *Grizzly Man*, Herzog says, "I wanted to make [Treadwell] what he was deep inside: one, a great performer, and for another, a great image-seeker" (qtd. in Garcia n.p.).
4. Some of the names, however, have more campy than familial origins. "I want to introduce you to one of the key role players in this year's expedition. The bear's name is The Grinch." Here a bear is not only given a name, but actually called a "role player" and happens to be named after a villainous Dr. Seuss character.
5. In the Discovery Channel special *The Grizzly Diaries* (2003), footage shows Treadwell engaged in a version of "fetch" with a fox, so a less reductive possibility—namely, the possibility that Treadwell and Ghost do actually *play* together—presents itself here.
6. Other footage not in Herzog's film suggests the extent to which Treadwell would go to enter the bears' world: the Discovery Channel's *Grizzly Diaries* (2003) shows Treadwell crawling on all fours miming the bears' snuffling along the ground, scratching his back on a pine tree, as well as his aping the grizzlies' clam-digging efforts along the beach. Treadwell would even sleep in the bears' own nest-beds (Jans 22). In one hypermimetic performance, when Treadwell was startled by humans in his wilderness sanctuary, he allegedly "dropped down on all fours and started making bear noises, similar to a female calling to her cubs, trying to warn them" (Jans 70).
7. In footage shown in Quinth's documentary *Deadly Passion*, Treadwell uses the term "disciplining" to characterize his behaviour toward unruly bears, which might indeed suggest a perceived fatherly role.
8. The forging of intimate relations by manipulating visual imagery is parodied delightfully in the web video series *The Guild* (2008), which centres on a group of friends who game together obsessively online. After falling in love with fellow guild member Codex (Felicia Day), who resists his advances, Zaboo (Sandeep Parikh) nevertheless assures her reciprocity (to his own satisfaction, at least) by splicing his own image into her DMV photo to make it appear that they are in fact "together."
9. Many viewers and critics chafed at being told by an on-camera Herzog how they should respond to scenes in *Grizzly Man*. As Dewberry notes, Carlo

Cavagna was not alone in complaining when he said that "Herzog comes across as the worst kind of cinematic Jackass—the filmmaker who doesn't trust his own work to speak for itself" (qtd. in Dewberry 2).

10. A less-charitable reading is that Herzog through the device of the headphones creates a "snuff film *sans image*" (Orgeron 59).

11. Note how the fox receives Timothy's own name for a title.

12. It might be worth considering to what genre or genres such "plotting" by Treadwell belongs as well. Herzog points out that Treadwell's footage goes "beyond a wildlife film." Treadwell often took fifteen takes to get his performance right, and as Dewberry puts it, "Herzog's film cuts to a scene where Treadwell emerges out of the Alaskan bush running toward the camera, as if staging an 'action-movie scene'" (2)

13. As Nick Jans points out in *The Grizzly Maze: Timothy Treadwell's Fatal Obsession with Alaskan Bears*, Treadwell was in fact named "Timothy Dexter" at birth (12). "Treadwell" is an assumed "stage name," yet another instance of his performative impulse in play. (Note moreover the wanderlust implied in one who "treads well." One finds echoes of this urge toward both celebrity and total freedom in Christopher McCandless's naming himself "Alexander Supertramp" in John Krakauer's book and Sean Penn's film *Into the Wild*.) Treadwell took his performance to extremes that exceeded even those of McCandless, however. After he moved out West in the 1980s, soon the name Dexter was gone and, as Jans puts it, "Timothy Treadwell, English waif, was born—complete with a Union Jack emblazoned on his surfboard and a Cockney accent so studied it fooled everyone, including the occasional Limey tourist and the English owner of the bar where he once worked" (13). He later reinvented himself as an Australian.

14. Some nuancing of nature as setting (Herzog) versus nature as character (Treadwell) is in order here. Even if Herzog's nature counts as setting, this setting is always much more for him than mere backdrop. For example, the opening sequence in *Aguirre: The Wrath of God* upset Kinski because it lacked any close-ups of his signature face; rather it was the face of the mountain, covered by tiny plodding human bodies, that dominated (Herzog, *My Best Fiend*). Such images exhibit nature's indomitable presence without regarding that presence as a personality. Treadwell's move to make nature a character might in turn count as more than mere personification, for he appears to consider his own footage as *revealing* rather than *ascribing* personalities to the bears. In this way he invokes a standard move in environmental literature to suggest that humans embody merely one way to exhibit personality along with countless other persons inhabiting a shared more-than-human drama. Crucially, Treadwell reveals bears as characters not only by making them seem like people, but also through his own ursine mimetics—crawling, digging, smearing himself with bear dung—which made him in turn seem more like a bear. Author Henry Williamson likewise waded rivers and crawled riverbanks in order to take on a mustelid's narrative perspective in *Tarka the Otter*. Chapter 1 of Aldo Leopold's *A Sand County Almanac* is a fine case study in how to transform a natural landscape from setting to character: The winter thaw evokes diverse responses from a host of different characters, including hawk, mouse, skunk—and human—each one viewing the melting snows from a concerned standpoint unique to that particular creature.

15. Of course, much depends on what counts as "speaking" and "listening" to wild nature here. Nick Jans relates a Tlingit man's advice for invoking both communication and kin-relationship with the grizzly while still respecting the grizzly's essential otherness: "If I'm going out to hunt or pick berries [. . .] I always do this: clap two or three times and say, 'Grandfather, I'm coming

into your woods. I won't stay long and I don't want to bother you.' Always let Grandfather know what you're up to, and he'll let you by" (188).

16. In the scene where Herzog interviews Treadwell's parents, it's worth noting that in addition to its depiction of a young Treadwell clutching his teddy bear, the family yard is itself filled with animal simulacra including ceramic bunnies and a fake goose, suggesting further artifice at work in the family's imagining of wild landscapes.

WORKS CITED

Abram, David. "Between the Body and the Breathing Earth: A Reply to Ted Toadvine." *Environmental Ethics* 27 (2005): 171–190. Print.

Aguirre: The Wrath of God. Dir. Werner Herzog. Werner Herzog Filmproduktion, 1972. Film.

Benjamin, Walter. "The Work of Art in the Age of Its Technological Reproducibility." *The Work of Art in the Age of its Technological Reproducibility, and Other Writings on Media.* Trans. Edmund Jephcott. Cambridge: Belknap Press of Harvard University Press, 2008. 19–55. Print.

Bolter, Jay and Richard Grusin. *Remediation: Understanding New Media.* Cambridge: MIT Press, 2000. Print.

Brautigan, Richard. "All Watched Over by Machines of Loving Grace." www.redhousebooks.com/galleries/freePoems/allWatchedOver.htm. Web. November 2013.

Buell, Lawrence. *The Environmental Imagination.* Cambridge: Belknap Press of Harvard University Press, 1995. Print.

Butler, Judith. *Gender Trouble.* New York: Routledge, 2006. Print.

Days of Heaven. Dir. Terrence Malick. Paramount Pictures, 1978. Film.

Deadly Passion: Tragedy in Katmai. Dir. Stefan Quinth. Camera Q, 2005. Film.

DeLuca, Kevin. Letter to the author. 29 September 2009. Print.

———. "Thinking With Heidegger: Rethinking Environmental Theory and Practice." *Ethics & the Environment* 10:1 (2005). Print.

Derrida, Jacques. *Of Grammatology.* Baltimore: Johns Hopkins University Press, 1998. Print.

Dewberry, Eric. "Conceiving *Grizzly Man* through the 'Powers of the False.'" *Scope* 11 (2008): 1–12. Print.

The Enigma of Kaspar Hauser. Dir. Werner Herzog. Werner Herzog Filmproduktion, 1974. Film.

Fata Morgana. Dir. Werner Herzog. Werner Herzog Filmproduktion, 1971. Film.

Faulkner, William. *The Bear. Six Great Modern Short Novels.* New York: Dell, 1954. 349–448. Print.

Fitzcarraldo. Dir. Werner Herzog. Werner Herzog Filmproduktion, 1982. Film.

Garcia, Maria. "Grizzly Tale: Werner Herzog Chronicles Life of Ill-Fated Naturalist." *Film Journal International* 108:8 (2005): 14–16. Print.

Gifford, Terry. *Pastoral.* New York: Routledge, 1999. Print.

Grizzly Man. Dir. Werner Herzog. Discovery Films, 2005. Film.

Grizzly Man Diaries. Dir. N/A. Discovery Channel, 2008. Film.

The Guild. Season 1, episode 3. Dirs. Jane Selle Morgan and Greg Benson. Independent, 2008. www.netflix.ca. Web. July 2014.

Haraway, Donna. *Simians, Cyborgs, and Women: The Reinvention of Nature.* New York: Routledge, 1991. Print.

Hayles, N. Katherine. "Saving the Subject: Remediation in House of Leaves." *American Literature* 74:4 (2002): 779–806. Print.

Heart of Glass. Dir. Werner Herzog. Werner Herzog Filmproduktion, 1976. Film.

Heidegger, Martin. *Basic Writings*. New York: Harper & Row, 1977. Print.

Herzog, Werner. "Minnesota Declaration: Truth and Fact in Documentary Cinema." www.wernerherzog.com/main/de/html/news/Minnesota_Declaration.htm. Web. November 2013.

Hirsch, Joshua. *Afterimage: Film, Trauma, and the Holocaust*. Philadelphia: Temple University Press, 2004. Print.

Into the Wild. Dir. Sean Penn. Square One, 2007. Film.

Jans, Nick. *The Grizzly Maze*. New York: Plume, 2006. Print.

Krakauer, Jon. *Into the Wild*. Garden City: Anchor Books, 2007. Print.

Leopold, Aldo. *A Sand County Almanac: And Sketches Here and There*. Oxford: Oxford University Press, 1949.

Marx, Leo. *The Machine in the Garden: Technology and the Pastoral Ideal in America*. Oxford: Oxford University Press, 2000. Print.

Orgeron, Marsha and Devin Orgeron. "Familial Pursuits, Editorial Acts: Documentaries after the Age of Home Video." *Velvet Light Trap* 60 (Fall 2007): 47–62. Print.

Peters, John Durham. "Kittler Seminar." University of Utah, October 2008. Speech.

Prager, Brad. *The Cinema of Werner Herzog: Aesthetic Ecstasy and Truth*. London: Wallflower Press, 2007. Print.

Rescue Dawn. Dir. Werner Herzog. Gibraltar Films and Thelma Productions, 2006. Film.

Rousseau, Jean-Jacques. *Confessions*. Trans. Angela Scholar. Oxford: Oxford University Press, 2000. Print.

Said, Edward. *Orientalism*. New York: Vintage, 1979. Print.

Sontag, Susan. "Notes on Camp." *Against Interpretation and Other Essays*. New York: Picador, 2001. 275–292. Print.

Stroszek. Dir. Werner Herzog. Werner Herzog Filmproduktion, 1976. Film.

Tatum, Stephen. *In the Remington Moment*. Lincoln: University of Nebraska Press, 2010. Print.

Where the Green Ants Dream. Dir. Werner Herzog. Werner Herzog Filmproduktion, 1984. Film.

Williamson, Henry. *Tarka the Otter*. London: Putnam and Sons, 1927. Print.

The White Diamond. Dir. Werner Herzog. Werner Herzog Filmproduktion, 2004. Film.

Wilde, Oscar. *The Picture of Dorian Gray*. London: Penguin, [1890] 2000. Print.

5 The Dangers of Biosecurity
The Host (2006) and the Geopolitics of Outbreak

Hsuan L. Hsu

The highest-grossing South Korean film in history, Bong Joon-ho's *The Host* [*Gwoemul*] (2006) garnered both widespread popularity and critical acclaim. The film is often regarded as either a South Korean version of a Hollywood monster movie, or a comic inversion of the conventions of traditional monster films. When it was released in the U.S. and other Western countries in 2007, *The Host* received rave reviews in venues ranging from *Rolling Stone* to *The New Yorker*. Many of these reviews focus on the film's computer-generated monster, comparing it to classic films like *King Kong* (1933), *Godzilla* (1954), and *Jaws* (1975); *New York Magazine* praised *The Host* as "one of the greatest monster movies ever made!" (Hill). However, Bong himself has described a more vexed relationship to traditional monster movies: "I have a real love and hate feeling towards American genre movies. I'll follow the genre conventions for a while, then I want to break out and turn them upside-down" ("'The Host'—Bong Joon-Ho Q & A").[1] Writing in *The New York Times*, Manohla Dargis attributes the film's originality to this dissolution of genre conventions: "'The Host' is a loose, almost borderline messy film, one that sometimes feels like a mash-up of contrasting, at times warring movies, methods and moods. Mr. Bong would as soon have us shriek with laughter as with fright. But it's precisely that looseness, that willingness to depart from the narrative straight and narrow, that makes the film feel closer to a new chapter than a retread" (Dargis n.p.).

As it unfolds through a series of digressions, fictional news clips, and multiple subplots, *The Host* combines generic conventions from monster movies, narratives of disease outbreak, news reportage, melodrama, and slapstick comedy. Early in the film, a gigantic amphibious creature emerges from the Han River, attacks dozens of bystanders, and kidnaps Hyun-seo, the daughter of the film's protagonist, Gang-du. Fearing that the creature carries a mysterious virus, the South Korean military steps in and quarantines Gang-du's entire family, along with anyone else who may have been exposed to the creature. When they learn that Hyun-seo is still alive, Gang-du, his two siblings, and his father escape from quarantine in order to rescue her. As the family scours the sewers of South Korea's capital city in search of the monster, the U.S. government and World Health Organization decide to circumvent the threat of an epidemic by treating the Han River area with an experimental biocidal chemical, Agent Yellow. Koreans organize a mass demonstration against the use of Agent Yellow, but it is deployed just as the family and the monster meet to fight it out. Following this final confrontation, Gang-du matures from a layabout to a responsible parent, assuming responsibility for an orphan who has been held captive in the creature's mouth. Throughout the film, Bong inserts passing references to politically charged events ranging from the U.S.-led "war on terror" and the hooded detainees at Abu Ghraib to the South Korean democratization protests of the 1980s and recent outbreaks of SARS and avian flu.

Without denying this film's resonances with classic monster movies, this chapter will argue that *The Host* can be productively interpreted as a revision of the popular epidemiological plot that Priscilla Wald has described as the "outbreak narrative"—a genre that legitimates Western scientific interventions and discourses of "development" while effectively blaming the results of underdevelopment on its victims. Through its intertwined genealogies of monstrosity, contagion, and biological hazard, *The Host* presents a critique of U.S. and international interventionism that stretches from the Korean War and the post-1997 structural adjustments imposed by the International Monetary Fund (IMF) to the biological and environmental harm caused by toxic dumping and chemical warfare. The film's focus quickly shifts from the amphibious creature—whose most spectacular exploits are over after the first twenty minutes of screen time—to the monstrous measures imposed by international interests more concerned with preserving the health of the population than with sustaining South Korea's capacities for social welfare and economic self-determination. The film thus shows how narratives of epidemiological outbreak mask the neoliberal economic reforms that have undermined traditional family life and the means of social reproduction associated with food, family support, and a healthy environment. Because *The Host* proceeds in the mode of pastiche—explicitly alluding to a range of popular films and genres that include *Jaws*, *Outbreak*, and news clips from the invasion of Iraq—this chapter will explore multiple points of origin for the film's numerous environmental, economic, evolutionary, and social embodiments of monstrosity. Yet all these themes are unified in the film's overarching treatment of the harmful effects of international military and economic interventions enacted in the name of security and public health.

TOXIC DEBT

> In another news report, we learn that the South Korean "industrial economy lauded by every U.S. President since Kennedy has mutated overnight into a nightmare of 'crony capitalism' in the twinkling of the I.M.F.'s eye" (Cumings 1998, 16). . . . [W]hat one reads in a novel about alcoholism and domestic abuse among construction

workers in Seoul can also be linked to the distant machinations on
Wall Street and in Washington.
—Amitava Kumar, "Introduction" to *World Bank*
Literature xviii–xix

Every fictional monster has its origins, and more often than not these lie in widespread anxieties about social and economic stability. As Annalee Newitz puts it in her study of monster narratives in U.S. popular culture, "the extreme horror we see in these stories—involving graphic depictions of death, mutilation, and mental anguish—is one way popular and literary fictions allegorize extremes of economic boom and bust . . ." (12). In its opening scenes, *The Host* offers multiple genealogies for its amphibious monster. The first of these is based on an incident that occurred in 2000, when Albert McFarland, the U.S. military mortician at the Yongsan camp, ordered two assistants to dump about 80 litters of formaldehyde into a sewage system that drains into the Han River. The incident outraged South Koreans, and has often been cited by environmental activists and demonstrators protesting against U.S. military presence. One *Korea Times* editorial, for example, notes that the Han River "supplies drinking water for over 10 million citizens," and highlights the symbolic violence inherent in this act, asking, "Are Koreans disposable people?" ("Editorial"). (McFarland's light punishment—a thirty-day suspension by the United States Forces Korea (USFK) and a $4,000 fine by South Korea's Ministry of Justice—indicates how little value was placed upon Korean citizens and their environment.)

The formaldehyde dumping scene in *The Host* emphasizes the American scientist's awareness of—and disdain for—the regulations he is violating. The first words of the movie—"Mr. Kim, I hate dust more than anything"— draw attention to the incommensurability between cleaning up the military morgue and dumping toxic waste in the river. The doctor orders a Korean assistant to dump the bottles of "dirty formaldehyde" because "every bottle is coated with layers of dust."[2] When the assistant protests that the chemicals will end up in the river, the mortician responds, "The Han River is very broad, Mr. Kim. Let's try to be broad-minded about this." The dialogue caricatures not only McFarland, but also the cynical discourse of liberal universalism that claims to be "broad-minded" while allowing and even endorsing the despoliation of vulnerable environments and populations.

The Host's widely noted anti-Americanism should also be situated in the context of ongoing demonstrations against U.S. military bases. On May 4, just before the July release of *Gwoemul* in South Korea, a demonstration protesting the expansion of a U.S. military base in the vicinity of Pyeongtaek was violently confronted by the South Korean military, who injured over two hundred people. In following weeks, demonstrations against the U.S. military presence spread; two thousand students marched from Seoul to Pyeongtaek, chanting "Yankees go home! This is our land!" (Persaud). Even the closures of many USFK bases were criticized by environmental groups, because South Korea accepted the closures despite eighteen months of debate about the environmental conditions of the sites (Slavin). In all these cases, the Korean government and military made clear concessions to the U.S., whether by suppressing protestors or by failing to exact adequate redress for environmental harm.

Immediately following the formaldehyde scene, Bong provides a second origin story for the mutated river monster at the centre of his film. Years after the toxic dumping incident, and a few months after a miniscule mutant creature bites the hand of a man fishing in the river, a businessman commits suicide by jumping from a bridge into the Han River. Situated between the appearance of the small creature a few months earlier and the emergence of the full-fledged beast in the following scenes, this suicide seems to play a key role in the monster's prodigious growth: The businessman may be the first human meal it eats. Furthermore, because Bong notes that such suicides in the Han River happen "almost every day," the monster's growth may be directly correlated to conditions affecting the Korean economy and those whose livelihoods depend upon it (Bong, "Audio Commentary"). South Korean newspapers in the last decade have referred to suicides by unemployed or bankrupt businessmen as "IMF suicides," linking the causes of their despair to the neoliberal structural adjustment program imposed by the IMF after the Asian Financial Crisis of 1997.[3] In light of this second origin story for Bong's monster, which links it to the epidemic of suicides resulting from South Korea's decimated economy, *The Host* turns out to be an allegory not just of American military occupation, but also of neoliberal market reforms. Bong is not alone in making films about the post-crisis economic situation: *Oldboy* (2003), directed by his friend Park Chan-Wook, dramatized the dismantling of Korean business conglomerates (*chaebŏls*) guided by the state and also included a digressive episode in which the protagonist encounters a suicidal businessman (see Jeon). Rob Wilson, in a study of carnal and violent eruptions of "killer capitalism" in post-9/11 Korean cinema, identifies Park as a leading practitioner of "'IMF-noir' [a term coined by Jin SuhJirn], reflecting tensions and phobias released in Korea, as across inter-Asia, by the monetary and fiduciary crisis of 1997, unmasking globalism" (127).

The liberalization of trade, labour markets, and investments imposed by the IMF in the wake of the 1997 crisis had devastating effects on South Korea's economy. State-guided corporations and banks (whose collective success prior to the crisis had been touted as the "Miracle on the Han River") were rapidly privatized, and businesses stayed afloat by cutting jobs. Economists James Crotty and Kang-Kook Lee attribute South Korea's social and economic instability to an influx of finance capital that has made "[t]he Korean stock market . . . a gambling casino for foreigners" interested in "short-term speculative profit rather than long-term growth" (671, 673). Sociologist Walden Bello adds that, since 1997,

[t]he IMF has touted Korea as a "success story." However, Koreans hate the Fund and point to the high social costs of the so-called success.

According to South Korean government figures, the proportion of the population living below the "minimum livelihood income"—a measure of the poverty rate—rose from 3.1 per cent in 1996 to 8.2 per cent in 2000 to 11.6 per cent in early 2006. The Gini coefficient that measures inequality jumped from 0.27 to 0.34. Social solidarity is unraveling, with emigration, family desertion, and divorce rising alarmingly, along with the skyrocketing suicide rate.

(Bello n.p.)

The nation's "inadequate" welfare system (Crotty and Lee 671), widespread unemployment, and unravelling social fabric suggest that, in economic as well as environmental terms, neoliberalism has transformed Korean citizens into "disposable people."[4] Austerity measures imposed by the IMF have not only decimated social services in South Korea, but also damaged its environment: "The ratio of the Ministry of Environment's budget to total finance fell from 1.51 percent in 1997 to 1.38 percent and 1.36 percent in 1998 and 1999, respectively. In order to attract foreign investment the South Korean government, on the advice of the IMF, has abolished or weakened various regulations. For example the government has removed the Green Belt Regulation, reorganised National Parks and weakened the regulations protecting sources of drinking water" (Chomthongdi). The businessman's suicide associates the film's overgrown, preternaturally mobile, computer-generated (by The Orphanage and Weta Digital, international firms based in California and New Zealand), voracious, flexible, and acrobatic beast with the post-crisis attrition of "social solidarity." The monster may have literally fed on the skyrocketing number of suicides precipitated by the dismantling of the pre-1997 state-guided economy.

These two originary scenes help account for the English-language title of Bong's film in international screening: whereas *Gwoemul* is best translated as "monster" or "creature," the much more ambiguous term "host" encompasses, among other things, questions about international hospitality. *The Host* blurs the distinction between the monster and the society it ravages. From its inception the movie asks, isn't South Korea playing host, arguably against the interests of its citizenry, to the U.S. military and to foreign investors who have no interest in the nation's social fabric? Despite its incompetence and palpable absence when it comes to confronting the monster, the Korean military plays an active role in enforcing a quarantine and suppressing demonstrations by its own citizens. Because the word "host" combines notions of hospitality, hostility, and armed forces, it significantly blurs the distinctions between the creature, the Korean army, and the nation's hospitality to foreign influences. *The Host* is not the first monster movie to allegorize economic and social ailments; however, the international scope of the film's satire calls for an analysis that triangulates between international interventions, the South Korean state, and the family of average (or below-average) Korean citizens who do battle with the monster. The following sections of this chapter examine how national and international responses to the threat posed by the monster are structured by a pursuit of "immunization" that neglects and hinders processes of social reproduction and thus exacerbates already existing risks of economic, environmental, and public health crises.

OUTBREAK AND EMERGENCY

> With the past decisions on nuclear energy and our contemporary decisions on the use of genetic technology, human genetics, nanotechnology, computer sciences and so forth, we set off unpredictable, uncontrollable and incommunicable consequences that endanger life on earth.
>
> —Ulrich Beck

While it invokes numerous associations with hospitality, guests, hostages, and hostility, the film's English-language title more directly draws attention to the creature as a suspected "host" or carrier of a fatal disease. Directly influenced by Hollywood films about biological threats,[5] this outbreak narrative—which erupts when an American sergeant who fought the monster develops a skin infection and the government declares the vicinity of the Han River a biohazard zone—shifts the film's plot from the conventional military confrontation with a monster to a more mediated and abstract struggle against a contagious disease. Rather than combating the disease vector—the monster itself—the state deploys medical specialists to examine citizens who may have had contact with the virus, and mobilizes the army to manage the movements and contacts of Seoul's inhabitants. By interpolating a satirical outbreak narrative into his monster movie, Bong suggests that discourses of biosecurity may constitute the real threat to the everyday well-being of South Korea's population (Figure 5.1).

In recent years, a range of biological and environmental threats have made the possibility of massive public health crises a prominent issue in debates about international security. In the wake of anxieties provoked by the AIDS crisis, viruses such as SARS, avian flu, mad cow disease, and West Nile raise the spectre of widespread epidemics of international proportions. Likewise, the threat supposedly posed by scenarios of bioterrorism involving anthrax, ricin, or emerging infectious diseases[6] has linked fears of biological epidemics to xenophobic anxieties about the porous borders of nations and bodies. Discourses and practices of biosecurity focus on containing risk and minimizing contagion in cases of dramatic outbreak, thus deemphasizing the various forms of structural violence that make particular groups and regions vulnerable to a range of maladies, both "natural" and artificially induced.

Discourses of biosecurity are at once nationalist in sensibility and transnational in scope. They encompass—and often pathologize—vulnerabilities in

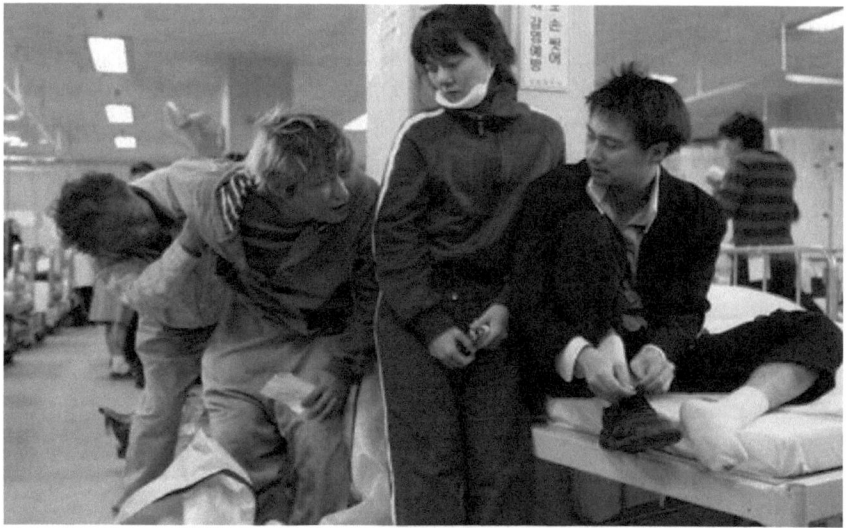

Figure 5.1 An unforgiving global health agency quarantines the central characters in *The Host*.

health, resources, and infrastructure produced by centuries of colonial and postcolonial exploitation. They often reproduce what Roberto Esposito has called an "immunization paradigm," which he opposes to the reciprocity of *communitas*: "We can say that generally *immunitas*, to the degree it protects the one who carries it from risky contact with those who lack it, restores its own borders that were jeopardized by the common" (27). Insofar as it "implies a substitution or an opposition of private or individualistic models with a form of communitary organization," immunization aptly describes the process by which states and international organizations protect themselves from "risky contact" with groups that have been rendered "risky" by the international community itself—by waves of colonization, occupation, war, and disinvestment that have decimated environmental, economic, and familial foundations for public health (27). At both the national and international scales, public health interventions too often "are shaped more by disease-oriented institutions of social control than by health oriented institutions of social justice" (Wing and Schinasi 790).

With the increase in public health concerns about emerging diseases, biosecurity has become a prominent theme in popular nonfiction, novels, and films. Genres that have been described as "biothrillers," the "killer virus novel," and the "outbreak narrative" have dramatized the horrifying threats posed by emerging diseases, the epidemiologically risky conditions of developing countries, and the heroic interventions of Western scientists in arresting the spread of contagion. Cinematic representations of emergent threats—which play a key role in reproducing "common-sense" anxieties

about biosecurity—can be traced from monster movies about nuclear proliferation, such as *Godzilla* (1954) and *Them!* (1954), to the anticipatory
resonances with the AIDS outbreak that Daniel Selden has identified in *Jaws*
(1975) and the Ebola scare dramatized in Wolfgang Petersen's *Outbreak*
(1995).[7]

Writing about U.S. representations of contagion, Priscilla Wald traces the
development, in both popular culture and medical discourse, of the "outbreak narrative," a genre that dramatizes the outbreak and eventual suppression of horrific biological threats. She describes

> the proliferation in the United States since the late 1980s of tales of con-
> tagious and infectious diseases emerging in Africa and posing a global
> threat until contained by dedicated—often maverick—public health
> officials and scientists in the United States whose triumphs allow them
> to reclaim modernity. The stories, which enable a displacement of the
> uncontainable and *domestic* threat of AIDS onto those infections, are
> therefore at least as reassuring to a Western audience as they are alarming.
> (691)

Citing popular films, fiction, and nonfiction such as *Outbreak*, Patrick
Lynch's *Carriers* (1996), and Richard Preston's *The Hot Zone* (1994), Wald
uncovers a metanarrative in which "an increasingly interconnected world
disturbs the lair of an archaic entity, a virus depicted as lying in wait, and
thereby brings modernity itself into conflict with a forgotten past, emblem-
atized by a disease against which contemporary technology is (initially)
ineffective: the return of a colonial repressed" (690–91). At stake in such
narratives is a pathologizing projection of vulnerabilities reproduced—if not
enabled in the first place—by modernization into foreign spaces, as well as
a misleading confidence that the solution to such crises in public health lies
in further biomedical research, rather than addressing economic and infra-
structural inequalities. For example, the notion that epidemics result from
a lack of exposure to Western hygiene, knowledge, and technologies diverts
attention from the extent to which the increasing incidence of emerging
diseases results from the expanded scope and mobility of biological vectors
caused by climate change or the increased vulnerability to disease caused by
the privatization of water sources and health care. In addition, such narratives downplay the resurgence of older, curable diseases such as cholera and
polio in nations deprived of resources by debt repayments and structural
adjustments imposed by the IMF and World Bank. Bong's awareness of the
international, economic sources of epidemiological vulnerability is evident
in a news clip that was edited out of the film. The clip features a WHO
(World Health Organization) website on avian influenza, with a highlighted
passage stating that "the next pandemic is likely to result in 57–132 million outpatient visits and 1.0–2.3 million hospitalizations, and 280,000–
650,000 deaths over less than two years. The impact of the next pandemic is

likely to be greatest in developing countries where health care resources are strained and the general population is weakened by poor health and nutrition" (Bong, "Deleted News Clips").

By exaggerating both the threat posed by emerging infectious diseases and the efficacy of Western science in treating them, outbreak narratives mask the economic motives behind many "global health" initiatives. In "Security, Disease, Commerce: Ideologies of Postcolonial Global Health," Nicholas King provides a lucid account of an "emerging diseases worldview" characterized by U.S. interests in surveilling and managing epidemiological risks throughout the developing world (767). King shows that discourses of biosecurity have emphasized not only humanitarian motives, but also the U.S.'s economic stakes in the health of people around the world. A 1997 report by the National Academy of Science's Institute of Medicine titled *America's Vital Interest in Global Health*, for example, notes that "America has a vital interest and direct stake in the health of people around the globe. . . . Our considered involvement can serve to protect our citizens, enhance our economy, and advance US interests abroad" (ctd in King 771). Thus, King notes that U.S. institutions like the Centers for Disease Control (CDC) and private investors in infrastructure and information services profit from a "global surveillance network" designed to identify and respond to epidemiological outbreaks. In the long run, the "emerging diseases worldview" bolstered by popular outbreak narratives attempts to distribute Western medical technologies and promote healthy populations "in an effort to foster the integration of underdeveloped nations into the world capitalist economy" (780). Perhaps the most catastrophic fault of such an approach, as Paul Farmer points out, is its blindness to the significant role of social and economic inequality in producing epidemiological vulnerabilities: "In its report on emerging infections," Farmer writes, "the Institute of Medicine lists neither poverty nor inequality as 'causes of [disease] emergence" (261–62).

The Host presents an incisive critique of the racial, colonial, and liberal presumptions that underlie the outbreak narrative. As the outbreak plot develops, a news program included in the film reports that the U.S. Centers for Disease Control analyzed tissue samples from an infected American officer and "confirmed that the creature from the Han River, as with the Chinese civet wildcat and SARS, is the host of this deadly new virus." The effects of toxicity have travelled full circle: from the bottles of formaldehyde in a U.S. military morgue to the body of an officer treated at a U.S. military hospital. Ironically, as the biosecurity emergency comes into effect in Seoul, we are shown a brief decontextualized news clip in which an American physician is stating, "I can't give any of that information without the approval of the United States." The emergence and characteristics of the virus are from the outset extraterritorial in nature: symptoms develop in the body of an American exposed to an entity created by American toxic waste, treated in a "U.S. military hospital," and further analyzed at the CDC. Details about the virus are U.S. state secrets: The mere knowledge that a virus exists suffices to legitimate a state of emergency and to manage the movements and fears of Seoul's citizens accordingly.

When Gang-du and his family escape from the hospital, we glimpse a news program reporting that the U.S. and other nations are concerned that "Korea is not adequately quarantining the infected"; even after Gang-du is recaptured (with a black hood labelled "Biohazard" placed over his head), we learn that "the U.S. and WHO, citing the failure of the Korean government to secure the remaining two family members or to capture the creature in question, have announced a policy of direct intervention." Further news reports—all of them accompanied by images that refer to campaigns in the Middle East, avian flu, and SARS—reveal the form that this intervention will take: "Agent Yellow, which has been chosen for use here in Korea, is a state-of-the-art chemical and deployment system recently developed by the U.S. to fight virus outbreaks or biological terror. This extremely powerful and effective system, once activated, completely annihilates all biological agents within a radius of dozens of kilometres." In Bong's parody of biosecurity measures, "biological terror" is combated not with the international sharing of research and vaccines,[8] but rather by killing every living thing in the endangered area: depending on how many "dozens of kilometres" are affected, the experimental deployment of Agent Yellow could potentially depopulate the entire city of Seoul. Of course, because the monster was created in the first place by a biocidal chemical (formaldehyde), this second dose of hazardous chemicals could potentially lead to further mutations and even more monstrous biological threats—not unlike the ongoing ecological and biological devastation produced by Agent Yellow's namesake, Agent Orange, in Vietnam.[9] The visual resemblance between the hanging yellow pod that delivers Agent Yellow and the pod-like shape of the monster underscores the circular logic behind this use of one chemical agent to fight the effects of another. As the film builds to its climax, the international community's plans to deploy Agent Yellow lead to a mass demonstration at the quarantined Wonhyo Bridge. The final battle between Gang-du's family and the monster is suggestively juxtaposed with the demonstration against foreign intervention in the name of biosecurity. Having so recently engaged in bitter democratization protests in the 1980s, Korean citizens now mobilize—both within and beyond the film—against international encroachments upon their sovereignty.

In a provocative critique of the U.S.'s shift towards a militarized, preemptive approach to dealing with emergent environmental and biological threats, Melinda Cooper argues that "social, biological, and environmental reproduction" are increasingly being viewed as matters of national security (92). Under the emerging agenda established during the Bush administration, war "is no longer waged in the defense of the state (the Schmittian philosophy of sovereign war) or even human life (humanitarian warfare; the human as bare life, according to Giorgio Agamben [1998]), but rather in *the name of life in its biospheric dimension, incorporating meteorology, epidemiology, and the evolution of all forms of life, from the microbe up*"

(98, emphasis in original). Yet, while experimental research addressing potential and as-yet nonexistent biological threats represents a promising site of capital investment, its efficacy in increasing, rather than undercutting, the security of "social, biological, and environmental reproduction" seems questionable. The biological research arm of the Pentagon, Cooper notes, "finds itself in the paradoxical situation of having first to create novel infectious agents or more virulent forms of existing pathogens in order to then engineer a cure" (91).[10]

While *The Host* dramatizes a similar scenario in which the possibility of a virulent biological agent leads to the intervention of state military and, subsequently, international forces, Bong's disease researchers do not "create" so much as they imagine and invent their emergent virus. In a chilling scene, Gang-du overhears an American scientist confiding to his translator that the American sergeant who had been in contact with the creature was not infected after all, but rather died from shock during his operation. Through a strange twist of logic that, as Tony Rayns suggests, resonates with the Bush administration's search for "weapons of mass destruction" (Bong, "Audio Commentary"), the doctor insists that, because "so far, there is no virus whatsoever," the virus must reside in Gang-du's head. Assisted by a team of Korean physicians, he then proceeds to violently probe Gang-du's brain for the virus. After superimposing a mysterious virus onto his monster movie plot, Bong leaves his viewers with an outbreak narrative without a virus. In the absence of a sensationalized biological threat to be dealt with by Western scientists in lab suits, we are forced to look elsewhere for virulent threats to biological well-being. In the following section, I will suggest that the dismantling of various means of social reproduction—including environmental integrity, family stability, and food security—poses just such a threat throughout *The Host*.

THE BELLY OF THE BEAST: SUBSISTENCE AND REPRODUCTION

> Social reproduction is the fleshy, messy, and indeterminate stuff of everyday life.
>
> —Cindi Katz

If biosecurity measures turn out to be the true perpetrators of "monstrosity" in Bong's film, the mutant creature itself turns out to have interesting, almost sympathetic qualities. For despite the many resemblances it bears to the ravages of the U.S. military and the IMF, the monster is more than a simple allegory of such incursions. Commenting on the film's climax in an interview, Bong confides that "when Gang-du . . . takes the pipe and strikes the monster in the mouth, I had a close-up of Gang-du and his face goes from rage to pity, as if he is thinking, 'You're sort of in the same situation as I am in.' It wasn't something that came as an accident. It was something that the actor and I discussed, that at this moment we should show some pity for the monster" (Bong, "The Han River"). In other words, Bong seems to have deliberately encoded into the film the notion that "[a]udiences taking in a monster story aren't horrified by the creature's otherness, but by its uncanny resemblance to ourselves" (Newitz 2). But in what ways can the creature and the humans it attacks be said to be "in the same situation"? Through a series of parallels, the film associates the creature not only with the devastating effects of toxic dumping, structural adjustment, and international intervention on South Korea's environment and economy, but also with basic subsistence activities necessary for the reproduction of life. At this creaturely level, the film's monster or "creature" turns out to share several qualities with those who are pursuing it.

If the creature turns out not to be a carrier of disease, it does play host to the two children, Hyun-seo and Se-ju, through much of the film. Bong has noted that, strictly speaking, this is "a kidnapping movie" structured by Hyun-seo's captivity ("Han River Horror Show"); but the creature's intentions for the children remain a mystery. Why does it swallow and regurgitate, without killing, each of the children during the course of the film? Why does it grab Hyun-seo with its tail as she is in the act of escaping, only to gently lower her to the floor? This may be explained as a simple necessity of the plot: The girl must be kidnapped and not killed so that her family can have a reason to escape from the hospital and hunt the monster. Why, then, would Bong introduce a second child into captivity, instead of stopping with Hyun-seo (Figure 5.2)? Se-ju's role in the film suggests that there is more than accident—or even the exigencies of plot—behind the children's survival.

Figure 5.2   The creature, with a young victim in tow, navigates Seoul's Han River, which the film suggests has been polluted through the country's industrial expansion.

Se-ju's appearance introduces one of the film's most noticeably digressive subplots. We first glimpse Se-ju and his older brother as they are hiding from Gang-du and his family, who are scouring Seoul's sewers in search of the monster. Without explanation, the camera follows the brothers away from the scene of the film's primary action, and we follow them to Gang-du's food stand in the quarantine zone. As they ransack the abandoned shop for food, Se-ju's brother explains that they are not stealing, but engaging in something "like melon *seo-ri* at a farm." As they leave the food stand, the brother explains that *seo-ri* is "an old borrowing game kids play. So *seo-ri* is a right of the hungry."[11] This definition abstracts from the rural "borrowing game" played by hungry children to a generalized and potentially revolutionary "right of the hungry" to appropriate or redistribute food. Ironically, immediately after this hungry street kid mentions *seo-ri*, the monster, exercising its own "right of the hungry," swoops down and swallows both brothers; when it regurgitates them in its sewer hideout, only Se-ju comes out alive.

The juxtaposition of the boys' and the creature's practices of urban foraging raises further questions about the monster's motives. Does it represent the monstrous threat posed by a "right of the hungry," or does the theme of *seo-ri* humanize the monster, which is just trying to satisfy its hunger? Does the creature's habit of depositing undigested (and in some cases still living) bodies in its hideout represent a tendency to hoard that violates the ethics of *seo-ri*, or does it demonstrate the creature's self-control in consuming no more than it needs? Such questions are further complicated when Gang-du's father, in order to excuse his son's apparent incompetence, explains that his son had been often neglected as a child: "And this poor boy with no mother . . . he must have been so hungry. Going around, doing *seo-ri* all the time. Raising himself on organic farms. Whenever he got caught, he'd get beaten up." Motherless, malnourished, and often sustaining blows to the head, Gang-du himself regularly practiced *seo-ri* to stay alive. He, too, seems intimately connected with the monster.

Interestingly, the description of Gang-du as a "poor boy with no mother" could easily apply to most of the film's major characters: Gang-du, his siblings, his daughter, and the two orphans subsisting in the sewers. There are no mothers in the film, and if (as Bong has suggested) "this weak family is in the middle of everything and the focus of the film," then the film is primarily concerned with the effects of the absence of maternal care (Bong, "Exclusive"). Early in the film, we learn that Hyun-seo's mother has abandoned her husband and daughter: "It's been 13 years since she popped out the baby and ran off. In a word, her birth was an accident. . . ."

The only creature in the film that pops out children and runs off is the monster: It swallows them alive, then delivers them to its lair through its intricate, fleshy mouth, which resembles a vulva—"a mouth that's a Freudian nightmare," in the words of one reviewer (Burr). The monster is implicitly compared with a mother, too, when (not yet having learned that Hyun-seo is still alive) Gang-du's father vows, "Until I slit that beast's stomach and at least find Hyun-seo's body, I'll never leave this world in peace." In this context, the creature appears to be an externalization of a model of motherhood gone awry—of the shortcomings of a "weak family" that fails to provide for social reproduction. In his influential study of *The Remasculinization of Korean Cinema*, Kyung Hyun Kim attributes the absence or marginalization of mothers in contemporary South Korean films to the country's rapid industrialization: "[F]renzied postwar urbanization had seriously altered familial relations to a point where 'mothers,' in their traditionally represented form, gradually disappeared from contemporary-milieu films" (6). The monster's maternal features thus reaffirm the connection to the IMF bailout I discussed earlier: Post-1997 neoliberal reforms led to social instability and rising rates of "emigration, family desertion, and divorce" (Bello), as well as a process of "casualization" or "flexibilization" of labour whose effects, according to Harvey, are "particularly deleterious for women" and therefore corrosive to households and families (112).[12]

On one level, then, fighting the monster is a way of punishing the "bad mother"—an externalization of the distressed functions of biological and social reproduction. In a sweeping analysis of "the monstrous-feminine" in horror films, Barbara Creed has argued that "when woman is represented as monstrous it is almost always in relation to her mothering and reproductive function" (7). On this view, battling the creature misogynistically compensates for anxieties about masculinity embodied by characters like the incompetent Gang-du, his unemployed alcoholic brother, and their father—a patriarch who dies after literally drawing a blank when attempting to shoot the creature with an empty shotgun. These crises of masculinity, in turn, register anxieties about economic failure (unemployment, families lacking support, the emigration of women and their employment in the informal economy). This reiterates a broader pattern of gender relations that Kyung Hyun Kim has observed in recent Korean films: "Through the relegation of the political crisis onto the body of a woman, the male subjectivities in a modern environment are born. The disfiguration of the woman covers up their incompetence and instability" (274). When Gang-du finally kills the creature by impaling its suggestively shaped mouth on a stick, the threat of undisciplined and excessively mobile motherhood is put down in no uncertain terms. Indeed, the climactic battle seems excessive, as the "weak" family, after recovering Hyun-seo's dead body from the monster's mouth, joins together in vindictively beating, shooting, burning, and impaling the mother-like creature, failing in their rage to register that the girl could just as easily have been killed by her exposure to a massive dose of Agent Yellow while the creature was carrying her. Gang-du's myopically vindictive focus on the monster as the cause of Hyun-seo's death effectively blinds him to the risks posed by Agent Yellow and international interventions in the name of public health. It also blinds him to the common capacities to feel hunger, pain, and hope that the creature shares with its human enemies and spectators.

The film's audience, by contrast, is invited to consider just these commonalities by two adjacent shots during the battle scene: first, a close shot of the creature's eye, already pierced by an flaming arrow shot by Gang-du's sister; next, a cut to the river, which the creature, half engulfed in flames, is presumably looking at in the hope that it will extinguish the fire. This creature's-eye-view of the Han River inverts the relation between the film's protagonists and its "monster" at the very moment when the creature is defeated, again raising questions about where the blame for Seoul's sufferings really lies

Whereas the battle with the monster seems excessively brutal, and (given all its correspondences with Gang-du and the practice of *seo-ri*) even senseless in its violent reassertion of manhood, the film's final scene offers an alternative resolution to the nation's broken families and ruptured social fabric. Gang-du fixes dinner for a sleeping boy, whom we recognize as Se-ju, the child befriended by Hyun-seo in the creature's hideout shortly before her death. Their shared meal echoes the film's only other scene of domestic harmony—a brief interlude in which Gang-du's entire family (Hyun-seo, still trapped in the sewer, makes a fictive or hallucinatory appearance) eat a simple meal together in their food stand, shortly before his father is killed by the creature. In both scenes, food and family sustain the characters just enough to remind us that, in *The Host*, both food and family are constantly threatened by the depressed economy, the monster, and state-sanctioned public health interventions. In the film's concluding meal—which takes place in the snack stand where the action began—there are two substitutions at work: the obvious substitution of Se-ju for Gang-du's lost daughter, and Gang-du's own transformation from a narcoleptic slacker into a responsible parent. Running the food stand, looking out the window for signs of monsters, and performing traditionally feminized domestic labour, Gang-du is inhabiting the roles of both his own dead father and the boy's absent mother.

The "weak family" at the centre of the film has been expanded and strengthened through an act of adoption—an act that is particularly significant given the cultural stigma[13] attached to domestic, single-parent adoption among many South Koreans, and in light of the effects of Korea's liberal transnational adoption policy upon the country's population, gender relations, and social wage. As Eleana Kim writes, "Adoptees and social activists in South Korea have criticized the state's continued reliance on international adoption as a social welfare policy solution . . . and its complicity in the perpetuation of gendered inequalities. Birth mothers—often working-class women, teen mothers, abandoned single mothers, sex workers, and victims of rape—represent the most subordinated groups in an entrenched patriarchy and misogynistic state welfare system . . ." (76).[14] In the case of transnational adoption, neoliberal reforms collude with traditional values (such as the stigma attached to adoption) to deprive both biological and potential adoptive parents of the capacity for reproduction. If the neoliberal economic depredations embodied allegorized by the monster have created prime conditions (such as financial, familial, and institutional instability) for the production of both real and "paper" orphans,[15] then Gang-du's adoption of Se-ju enacts a defiance of those conditions and a refusal to allow the child to either starve or be adopted internationally. As in its initial formaldehyde-dumping scene, the film's conclusion is also grounded in recent history: Just before *The Host* was released, the Korean Ministry of Health and Welfare announced that, in order to bolster domestic adoptions, it would allow—and provide financial incentives for—adoption by single-parent households (Park n.p.).

The Host thus concludes with a mundane yet unconventional act of adoption that to some extent restores Gang-du's and Se-ju's lost family ties. As an alternative to the vindictive assault on the creature in previous scenes, this quiet epilogue asserts that reproductive labour—cooking, housework, and child-rearing—plays an important role in the maintenance of life. Although Bong's film—like so many recent Korean films—suffers from a palpable absence of maternal characters, its ending gestures towards progressive domestic conditions that, according to Seungsook Moon, would be necessary for the reformation of South Korea's masculinized public sphere: "As long as the domestic identities of women as mothers/wives/daughters-in-law mediate women's citizenship, women's access to civil society is practically and ideologically hampered. . . . [Q]uotidian practices of housework, childrearing and extended family obligation primarily performed by women overshadow their citizenship rights that formal law is supposed to guarantee" (138). The spectacle of Gang-du and Se-ju sitting down to a home-cooked meal with the television turned off (the first meal we see that does not consist of processed foods like canned octopus or instant noodles) also offers an alternative interpretation of "biosecurity" in terms of social ties and reproductive work that require state support. For the risk factors that are at once named and misrepresented by the designation "biohazard" would be reduced by a more equitable distribution of the means of social reproduction: health care, welfare, and social services; access to food; and an environment supportive of life.[16] Thus, in an article that faults public health preparedness programs for overemphasizing the medical treatment of diseases, Steve Wing and Leah Schinasi stress that "[h]ealth care, as in 'care of health,' involves access to healthy foods, clean water, clean air, safe housing, rewarding jobs with livable wages, safe transportation, quality education, opportunities for physical and mental activity, and services such as medical services, as well as prevention of and protection from violence" (791).

In his influential report on Ebola, "Crisis in the Hot Zone" (1992), Richard Preston cites head of the National Institutes of Health Stephen Morse describing a scenario wherein an emerging disease could wipe out humankind. Morse explains that the genetic diversity of the population would prevent a virus from extinguishing the species, but "if one in three people on earth were killed—something like the Black Death in the Middle Ages—the breakdown of social organization could be just as deadly, almost a species-threatening event" (81). "Social organization"—which involves social reproduction as well as political and economic stability—turns out to be more vital than any emerging biological threat. For this reason, Ulrich Beck warns that neoliberalism's approach to managing "risk" is perilous in its decimation of public institutions designed to support social reproduction and civil society: "[T]he separation of the world economy from politics is illusionary. There is no security without the state and public service" (12). In its diegetic vacillations between monster plot, outbreak narrative, and a few mundane scenes of cooking, eating, and *seo-ri*, *The Host* bitterly criticizes the ways in which biosecurity discourses mask the harmful, often fatal, effects of economic neoliberalism. Whether the monster ultimately allegorizes toxic dumping, U.S. military occupation, the IMF, or the CDC, its assault on the weakened, motherless family abandoned by the state dramatizes the devastating effects of all these phenomena on South Korean social reproduction.

Less than two years after *The Host* was first released, the vital issues of biosecurity and food safety resurfaced in massive demonstrations in Seoul and other cities. The issue in May and June of 2008 was President Lee Myung-bak's promise to George Bush to resume U.S. beef imports, which had been banned since 2003 and which Koreans widely associate with BSE, or mad cow disease. Charles Armstrong reports that by June, there were "almost nightly candlelit protests in the centre of Seoul and other cities, estimated to have mobilized over a million Koreans" (116). Many of these demonstrators were "women who were extremely upset that in years to come their children might pay with their lives for President Lee's kowtowing to US export interests" (Hudson n.p.). Demonstrators expressed their concerns that mad cow disease might not have been purged from U.S. cattle, noting that "mad cow disease can remain dormant for decades in humans who have eaten tainted meat" (Hudson n.p.). In a striking rhetorical inversion of the outbreak narrative, South Koreans represented the U.S. as a source of contagion and technologically produced emergent risks. In this instance, biosecurity measures were invoked not to undermine national sovereignty, but to shore it up, and to criticize Lee's economic policies. While beef imports were the most proximate cause of the demonstrations, protests also spread to encompass larger issues such as "rising fuel prices[,] large-scale privatizations, rising education costs, [and] attacks on labour rights" during the first months of Lee's presidency (Armstrong 116). Like the monster attacks dramatized in *The Host*, these recent protests demonstrate the extent to which the outbreak narrative is intertwined with issues of national identity, economic neoliberalization, political sovereignty, and everyday practices of social reproduction

ACKNOWLEDGEMENTS

Thanks to Shameem Black, Joseph Jeon, Mark Jerng, Martha Lincoln, and Edlie Wong for generously comments on earlier drafts of this chapter.

NOTES

1. In "Globalisation and New Korean Cinema," Shin argues that this hybridizing deployment of Hollywood conventions and other international genres is characteristic of contemporary South Korean cinema, and that this enables the films to respond critically to both national and transnational issues: "These hybrid cultural forms provide an important means for . . . self-definition, a self-definition that not only distances itself from a xenophobic and moralizing adherence to local cultural 'tradition' but also challenges Western cultural hegemony" (57).
2. Although the exchange about dusty bottles seems far-fetched, Bong notes in an interview that it was based on McFarland's "real" reason for having the formaldehyde dumped (Bong, "Audio Commentary").
3. A longer account of the suicide that has been edited out of the film directly adduces the victim's bankruptcy and exorbitant credit card debt as motives for his death (Bong, "Deleted News Clips").
4. David Harvey provides a useful historical account of economic neoliberalization in South Korea, arguing that it did not create new wealth so much as it unevenly redistributed already existing wealth (106–12). Naomi Klein describes the IMF interventions in South Korea as a prominent example of how "disaster capitalism" forcibly privatizes public and state resources in the wake of (in this case economic) crisis (263–80). See also Bruce Cumings, "The Korean Crisis and the End of 'Late' Development."
5. Bong notes that the biohazard "costumes" and equipment that appear in the fabricated news clips are of the sort that "you can see in a movie like *Outbreak*" ("Audio Commentary").
6. In *Life as Surplus: Biotechnology and Capitalism in the Neoliberal Era*, Melinda Cooper notes that "[u]sing the technique of DNA shuffling (hailed as the second generation of genetic engineering because of its highly accelerated capacity for randomly recombining whole segments of genomes), DARPA [the Defense Advanced Research Projects Agency] is attempting not only to perfect our defenses against existing threats but more ambitiously to create antibiotics and vaccines against infectious diseases *that have not yet even emerged*" (91, emphasis in original).
7. See Mayer's analysis of "virus discourse" in the contemporary genre of the "biothriller," Dougherty on the "killer virus novel," and Wald's on the "outbreak narrative." Mayer points out that, as a discourse organizing national security, the virus reflects an ambivalence towards the "versatility" of global economic relations: "The virus, which may work its way from species to species through contaminated secretions or excretions and which is capable of changing the genetic material it comes into contact with, attests to a protean versatility that is further emphasized once the ambivalent nature of the pathogen—between life and death—comes into view. This ambivalence turns the virus into a perfect trope to envision contemporary world-political developments and interactions" (7). On *Jaws* as an anticipatory narrative of public reactions to AIDS, see Selden, "Just When You Thought It Was Safe to Go Back in the Water . . ."
8. On developed nations' failure to adequately share vaccines during outbreaks of avian influenza in underdeveloped nations, see Davis, 151–63.
9. While the name of this chemical substance clearly echoes the herbicide Agent Orange, it also invokes the yellow dust from the Gobi desert that has on several occasions brought carcinogenic pollutants from China's industrial cities into Seoul.

10. In March 2005, more than 750 microbiologists sent a letter to the director of the NIH in which they stated that the current funding, which prioritized biodefense over projects with public health significance, represented a threat to their field and hindered scientific progress (NIH). See also Nicholas B. King, "The Ethics of Biodefense."

11. Bong elaborates further, saying that *seo-ri* is "different from shoplifting or any kind of robbery—it's some kind of playing, game, it's some kind of rural culture of Korea: young boys in the night invade a fruit field or some place and take some fruit or sometimes even chickens. . . . It's some kind of culture of the lower class people . . ." (Bong, "Audio Commentary"). The theme of *seo-ri* is present from the film's first shot of Gang-du: He has fallen asleep while working at the snack booth, and a boy (Se-ju, as it turns out) attempts but fails to steal a piece of candy.

12. The *chaebŏl*, state-led economic system that came to an end with the economic crisis and IMF bailout also relied on extended "familial" ties between salarymen and their employers. "By cloaking the demands of profit in the guise of older Confucian practices, labor itself gets recoded as the work of the family and by extension of the nation" (Jeon).

13. "In South Korea, adoption is still not viewed as a socially acceptable alternative method of extending one's family or parenting a child" (Sarri, Baik, and Bombyk 100).

14. Sarri, Baik, and Bombyk provide a comprehensive analysis of the detrimental effects of transnational adoption on South Korea's welfare system and social programs for children and single mothers.

15. "An eleven-year decline in transnational South Korean adoption was reversed with the IMF crisis, which caused a concomitant crisis of overflowing orphanages. In 1996, approximately five thousand children were placed in state care, and that figure was projected to be double in 1998, leading the Ministry of Health and Welfare to announce that it 'has no choice but to make changes to recent policy which sought to restrict the number of children adopted overseas'" (E. Kim 64).

16. Social reproduction, Katz writes, "hinges upon the biological reproduction of the labor force, both generationally and on a daily basis, through the acquisition and distribution of the means of existence, including food, shelter, clothing, and health care" (710). Along with cultural and political-economic issues, Katz includes environmental harm as a factor in social reproduction: In both environmental racism and migrant labour patterns, "there is a rejigging of the geography of social reproduction so that the costs of social reproduction—in once case environmental and in the other political-economic—are borne away from where most of the benefits accrue" (714).

WORKS CITED

Armstrong, Charles. "Contesting the Peninsula." *New Left Review* 51 (2008): 115–35. Print.

Beck, Ulrich. "The Silence of Words and Political Dynamics in the World Risk Society." Trans. Elena Mancini. *Logos* 1:4 (2002): 1–18. Print.

Bello, Walden. "All Fall Down." *Foreign Policy in Focus* 3 July 2007. http://fpif.org/all_fall_down/. Web. July 2014.

Bong Joon-ho. "Audio Commentary." *The Host: Collector's Edition*. New York: Magnolia Home Entertainment, 2007. Film.

————. "Deleted News Clips." *The Host: Collector's Edition*. New York: Magnolia Home Entertainment, 2007. Film.

————. "Exclusive: *The Host*'s Bong Joon-ho." Interview with Edward Douglas. 6 March 2007. www.comingsoon.net/news/movienews.php?id=19126. Web. November 2013.

————. "The Han River Horror Show: An Interview with Bong Joon-ho." Interview with Kevin B. Lee. Trans. Ina Park and Mina Park. *Cineaste* 32:2 (2007). www.cineaste.com/articles/an-interview-with-bong-joon-ho.htm. Web. November 2013.

————. "The Host—Bong Joon-Ho Q & A." *Time Out New York* 7 November 2006. www.timeout.com/film/newyork/news/1514/the-host-bong-joon-ho-q-a.html. Web. November 2013.

Burr, Ty. "A Monster Movie for the 21st Century." *Boston.com* 9 March 2007. www.boston.com/movies/display?display=movie&id=9483. Web. November 2013.

Chomthongdi, Jacques-Chai. "The IMF's Asian Legacy." *Focus on Trade* September 2000. http://focusweb.org/publications/2000/The%20IMFs%20Asian%20Legacy.htm. Web. July 2014.

Cooper, Melinda. *Life as Surplus: Biotechnology and Capitalism in the Neoliberal Era*. Seattle: University of Washington Press, 2008. Print.

Creed, Barbara. *The Monstrous-Feminine: Film, Feminism, Psychoanalysis*. New York: Routledge, 1993. Print.

Crotty, James and Kang-Kook Lee. "The Effects of Neoliberal 'Reforms' on the Post-Crisis Korean Economy." *Review of Radical Political Economics* 38:4 (2006): 669–75. Print.

Cumings, Bruce. "The Korean Crisis and the End of 'Late' Development." *New Left Review* 231 (1998): 43–72. Print.

Dargis, Manohla. "It Came from the River, Hungry for Humans (Burp)." *New York Times* 9 March 2007. www.nytimes.com/2007/03/09/movies/09host.html. Web. November 2013.

Davis, Mike. *The Monster at Our Door: The Global Threat of Avian Flu*. New York: New Press, 2005. Print.

Dougherty, Stephen. "The Biopolitics of the Killer Virus Novel." *Cultural Critique* 48 (2001): 1–29. Print.

"Editorial: Are Koreans Disposable People?" *Korea Times* 17 July 2000. www.highbeam.com/doc/1G1-63489933.html. Web. November 2013.

Esposito, Roberto. "The Immunization Paradigm." Trans. Timothy Campbell. *diacritics* 36:2 (2006): 23–48. Print.

Farmer, Paul. "Social Inequalities and Emerging Infectious Diseases." *Emerging Infectious Diseases* 2:4 (1996): 259–69. Print.

Harvey, David. *A Brief History of Neoliberalism*. New York: Oxford University Press, 2007. Print.

Hill, Logan. "Three Steps: The New York Film Festival." *New York Magazine* 1 October 2006. http://nymag.com/movies/filmfestivals/21981/. Web. November 2013.

The Host: Collector's Edition. Dir. Bong Joon-Ho. Magnolia Home Entertainment, 2007. Film.

Hudson, Gavin. "Mad Cow Disease Fears Spark Mass Demonstrations in South Korea." *Ecoworldly.com* 8 May 2008. http://ecolocalizer.com/2008/05/08/mad-cow-disease-fears-cause-mass-demonstrations-in-south-korea/. Web. November 2013.

Jeon, Joseph Jonghyun. "Residual Selves: Trauma and Forgetting in Park Chan-wook's *Oldboy*." *Positions: East Asia Cultures Critique* 17:3 (2009): 713–40. Print.

Katz, Cindi. "Vagabond Capitalism and the Necessity of Social Reproduction." *Antipode* 33:4 (2001): 709–28. Print.

Kim, Eleana. "Wedding Citizenship and Culture: Korean Adoptees and the Global Family of Korea." *Social Text* 21:1 (2003): 57–81. Print.

Kim, Kyung Hyun. *The Remasculinization of Korean Cinema.* Durham: Duke University Press, 2004. Print.

King, Nicholas B. "The Ethics of Biodefense." *Bioethics* 19:4 (2005): 432–46. Print.

———. "Security, Disease, Commerce: Ideologies of Postcolonial Global Health." *Social Studies of Science* 32:5–6 (2002): 763–89. Print.

Klein, Naomi. *The Shock Doctrine: The Rise of Disaster Capitalism.* New York: Metropolitan Books, 2007. Print.

Kumar, Amitava. "Introduction." In *World Bank Literature*, xvii–xxxiii. Ed. Amitava Kumar. Minneapolis: University of Minnesota Press, 2003. Print.

Mayer, Ruth. "Virus Discourse: The Rhetoric of Threat and Terrorism in the Biothriller." *Cultural Critique* 66 (2007): 1–20. Print.

Moon, Seungsook. "Women and Civil Society in South Korea." In *Korean Society: Civil Society, Democracy, and the State*, 121–43. Second Edition. Ed. Charles K. Armstrong. New York: Routledge, 2002. Print.

National Institutes of Health (NIH), Office of Communications and Public Liaison, National Institute of Allergy and Infectious Diseases. "Open Letter in Science Regarding NIH Biodefense Funding." 17 March 2005. www.niaid.nih.gov/news/newsreleases/Documents/ebright_qas.pdf. Web. Accessed July 2014.

Newitz, Annalee. *Pretend We're Dead: Capitalist Monsters in American Pop Culture.* Durham: Duke University Press, 2006. Print.

Park Chung-a. "Singles Can Adopt Children." *Korea Times* 18 July 18 2006. http://times.hankooki.com/lpage/200607/kt2006071817382810160.htm. Web. November 2013.

Persaud, Natasha. "U.S. Base Expansion in Korea Sparks Protests." *Socialism and Liberation* August 2006. http://socialismandliberation.org/mag/index.php?aid=663. Web. November 2013.

Preston, Richard. "Crisis in the Hot Zone." *New Yorker* 26 October 1991: 58–81. Print.

Sarri, Rosemary C., Yenoak Baik and Marti Bombyk. "Goal Displacement and Dependency in South Korean-United States Intercountry Adoption." *Children and Youth Services Review* 20:2–1 (1998): 87–114. Print.

Selden, Daniel. "Just When You Thought It Was Safe to Go Back in the Water . . ." In *The Lesbian and Gay Studies Reader*, 221–26. Ed. Henry von Abelove, Michele Aina Barale, David M. Halperin. New York: Routledge, 1993. Print.

Shin, Jeeyoung. "Globalisation and New Korean Cinema." In *New Korean Cinema*, 51–62. Ed. Chi-Yun Shin and Julian Stringer. Edinburgh: Edinburgh University Press, 2005. Print.

Slavin, Eric. "South Korea Shelves Pollution Issue, Accepts 15 U.S. Sites." *Stars and Stripes* 17 July 2006. www2.pslweb.org/site/News2?page=NewsArticle&id=105 23&news_iv_ctrl=1701. Web. November 2013.

Wald, Priscilla. "Future Perfect: Grammar, Genes, and Geography." *New Literary History* 31 (2000): 681–708. Print.

Wilson, Rob. "Killer Capitalism on the Pacific Rim: Theorizing Major and Minor Modes of the Korean Global." *Boundary 2* 34:1 (2007): 115–33. Print.

Wing, Steve and Leah Schinasi. "Public Health Preparedness: Social Control or Social Justice?" *South Atlantic Quarterly* 106:4 (2007): 789–804. Print.

6 Biting Back
America, Nature, and Feminism in *Teeth*

Roland Finger

Written and directed by Mitchell Lichtenstein, *Teeth* (2007) gives biting social and environmentalist commentary. The satiric film shocks with its naïve, young Christian protagonist, Dawn, a girl next door who discovers and comes to grips with her genetic mutation, a powerful vagina dentata.[1] *Teeth* suggestively builds on an American tradition of associating women with nature, a theme that Annette Kolodny's *The Lay of the Land* masterfully explores within American culture. Kolodny shows how male explorers, settlers, and even flower children imagined America as feminine to satisfy their particular political and personal goals, ranging from the conquest of valuable land to the establishment of fraternal communities on nature's bosom. Crucially, the male fantasy about nature is double sided. It contains an impulse to master and possess nature, supposedly expanding the male realm of civilization, but the fantasy also reveals a male desire to be possessed by feminine nature and be relieved of the burden of a busy, harried life. This male desire to escape from civilization's anxieties is, as Kolodny puts it, "regression from the cares of adult life and a return to the primal warmth of womb or breast in a feminine landscape" (6). Men imagine themselves to represent civilization, rationality, and mastery spreading across America's land, fulfilling the narrative of manifest destiny, while they paradoxically also long to find carefree bliss and satisfaction acquired through immersion in feminized nature. Men dream of having it all: civilized dominion but also passionate release, available through the wonder lay of the American landscape. But Lichtenstein's protagonist, Dawn, is a new figure in the American tradition, an independent female pioneer who bites off narratives of extreme male authority. She is a modern mythical American figure, combining a woman's self-defence with nature's.

If women are classically identified with nature on whom men impose themselves for gain and pleasure, *Teeth* transforms the notion of feminine nature as merely attractive, passive, or nurturing into a self-preserving force, operating beyond the realm of typical consciousness. Lichtenstein modernizes vagina dentata myths to suggest that male abuse of feminine nature will no longer be tolerated; Dawn's inner power reflects nature's response to phallic dominance, the belief in male primacy and authority within our society. Whether men behave with ersatz political correctness or exploitative abandon, their quest for domination will not go unpunished. Dawn in her development as a vaginal vigilante—or a vagilante, as it were—rejects patriarchal tyranny, not initially through an explicit thought process but rather on a physical and unconscious level—on the level of the adaptable natural body. In Lacanian terms, Dawn as a figure of nature unconsciously attacks the phallus—which Gayle Rubin points out is "the [symbolic] embodiment of male status . . . an expression of the transmission of male dominance" (192). Lichtenstein mocks the idea of the phallus as a sign of power in our culture through Dawn's natural self-defence system. Lichtenstein's use of the vagina dentata myth spells out a bold critique of gendered social and environmental problems in contemporary America.

Set in an anywhere-in-America suburban landscape, *Teeth* tellingly foregrounds a towering phallic nuclear power plant, suggesting a prevalent masculine culture overlooking a feminized landscape. From the film's opening shot panning across a clean and green suburbia, we see that all is not right in this American garden where erect smokestacks pour out (or perhaps even ejaculate) pollution that grows more ominous as the film progresses. The phallic presence of the nuclear plant connotes that "the power" behind this society has been male, and it has mortally harmed, at least in one prominent case, an important woman in Dawn's life, her mother, Kim O'Keefe, whose name recalls Georgia O'Keefe and her art that often focuses on beautiful and delicate vaginal imagery within nature. While the phallic nuclear plant appears to be the cause of the cancer that afflicts Kim O'Keefe, her death is accelerated by the cold-hearted behaviour of her stepson, Brad, who is too obsessed with anally penetrating his girlfriend to heed his stepmother's cries for help as her condition becomes critical. Dawn is no victim like her dying mother. She is nature's response to nuclear pollution and violent actions associated with masculinity. Dawn represents a generation that has adapted to face modern gender problems.

The civil planner who would place a nuclear power plant in a suburban neighbourhood would have to be a cruel jokester, one who wants to ram home the idea that no landscape or population is safe from modern destructive phallic power. Even though Kim is the victim of nuclear pollution, her stepson's toxic behaviour is akin to nuclear waste. A raging sexist, Brad will only have anal sex with his girlfriend because of his phobia of vaginas, resulting from his childhood attempt to molest Dawn when she was approximately four or five years old and he was around eight or nine. When Brad tries to insert his index finger into Dawn's vagina, the little girl lacerates the tip of the offending digit with her vaginal teeth. Brad's attempt to molest his soon-to-be stepsister definitely forebodes his ill social adjustment that climaxes in his young adult life; he is apparently an unemployed, out-of-school leech, incapable of sincere affection or respect for his parents. He seems to spend all his free time (of which there is no other kind for Brad) playing computer games, smoking marijuana, and degrading his girlfriend.

He names his Rottweiler "Mother," apparently suggesting that he considers women to be subordinate dogs open to ridicule. Brad enjoys being the master of Mother, and we would probably not be too surprised to find out that Brad is "Mother/Dog Fucker." Brad is a cesspool of macho and sexist stereotypes.

Dawn is a breath of fresh spring air compared to her insensitive stoner stepbrother. Placing much faith in a Christian abstinence society called The Promise, Dawn depends upon this religious group to suppress and shackle the natural impulses of young people. A boy to whom she is attracted, Tobey, infiltrates The Promise group and tries to develop a relationship with Dawn. The portrayal of clumsily budding sexuality is meant to seem sweet, at first, as we see the young couple attending a G-rated film and later swimming in an idyllic blue lagoon setting. Dawn is gently experimenting with sexuality, but she does quickly put the brakes on Tobey's advances. He rubs her breasts as they climb up a rope swing at the lagoon, and she looks at him seriously and repeats, "Purity." Her response means please do not touch me in that way, and he temporarily pretends to respect her wishes. Behind the catchphrase "Purity," derived from the abstinence group, we recognize that Dawn stands against "pollution," which represents a combination of moral and physical corruption. Within the film, moral problems always correspond with physical realities. Brutal sexuality, unfeeling action, and nuclear waste, they are all of a piece and are associated with the phallic pollution of women and nature.

The picturesque lagoon is a temporary paradise where young lovers may explore their feelings for each other. It's a liminal space, not directly under the eye of society, where Dawn and Tobey allow some of their instincts to surface. In this area, the camera lingers on feminine nature images: a tree whorl shaped like a vagina, a cave gently veiled by a waterfall. The nature images provide gentle feminine relief before the onset of an insensitive phallic attack, and perhaps they suggest that Tobey is overly aroused by awareness of a yonic presence. The cave into which the young couple moves allows for confessions and the awakening of natural impulses; it's a place that suggests entry into a vagina, and it seems to be ready for safe exploration of desire, but the boyfriend goes too far. Tobey expects Dawn to give him sex, and before he date-rapes her, he implies that she has to comply because he has not masturbated since Easter. What exactly does this mean? Lichtenstein is certainly ridiculing Tobey's religious devotion, commenting that he does not really take the Christ rebirth story seriously, but also Lichtenstein seems to be emphasizing that Easter connects to pagan rituals of fertility, and so perhaps it is comically natural for a teenager like Tobey to masturbate on this particular day. We are probably not meant to imagine that Tobey gets hyper-stimulated when his family congregates for a holiday—not meant to think that there may be something about Grandma's holiday ham that awakens Tobey's hormones. Moreover, Tobey gives the impression that he has recently refused to masturbate because of Dawn's devotion to celibacy. Feeling that he has played Dawn's game long enough, Tobey now requires her to play his. The rape scene begins Dawn's development as a vagilante, a woman naturally resistant to abuse. The authority of the abstinence program becomes ludicrous before Tobey's violence. He knocks Dawn unconscious and penetrates her, but Dawn reacts instinctively to thwart him, nature and womanhood relying upon hidden inner resources. The film implies that men may attempt to rape nature, but eventually, the damage will boomerang. Mother Nature, of whom Dawn is a representative, will continue to adapt and thrive long after men endanger their own existence through short-sighted, selfish actions.

Barbara Creed points out in *The Monstrous-Feminine* that women have frequently been associated with nature because of birth imagery and menses; often, the feminine is portrayed as something animalistic or abject—something beyond the control of orderly, clean, and proper civilization. This kind of abjection, Creed notes, "is constructed as a rebellion of filthy, lustful, carnal, female flesh" (38). Many archetypes of males stand in antithesis to women as unfettered nature. Tapping into this tradition of gender contrast, Hawthorne's "The Birthmark" portrays Aylmer, a cerebral scientist and aesthete who decides to remove a tiny hand-shaped birthmark from the cheek of his wife, Georgiana. He decides that he wishes to purify what nature has placed upon her. His activity reveals an urge to attain a deeper level of possession and mastery over the natural feminine body. Aylmer views the birthmark as a taint of the flesh that he cannot stand, a blemish that overtly smacks of birth and sexuality—reminders of earthy bodily life that he wants to deny. While Georgiana is a victim with whom we sympathize, Dawn gains our sympathy by turning into a rebel against the status quo of phallic dominance.

Within *Teeth*, Dawn is not a figure of abjection. She does not bleed or become bloody as a result of her vaginal superpower. Men become the bleeders in the film and take on the burden of abject fleshiness. Their penile stumps quiver and shoot blood, as if they are ejaculating, but Dawn is never splattered. When Dawn does take a shower after she has been raped, she is not the damaged Marion Crane of Hitchcock's *Psycho* who dies just before she can pull her life back together, nor is she the knife wielding "mother," Mrs. Bates, who takes over the psyche of a weak man, Norman. Dawn is her own person, a castrator with a moral purpose.

Dawn is not heartless as she discovers her power. She sympathizes with Tobey, even though he has raped her. When she bites off his penis and he howls with confusion, she looks worried for him. Later, she is puzzled and horrified to see his penis being consumed by small crabs. This scene shows that part of her, the non-sentimental side, is very much like the crab following survival instincts. She reflects the practicality of nature that can at times seem to be cruel, but really she does what she must, and there is no moral culpability for her. But Dawn still feels bad for Tobey. After realizing that Tobey has bled to death, Dawn is transformed and rejects her adherence to orthodox Christianity, casting off her Promise Society ring from a cliff toward the blue lagoon. While she accepts lost sexual innocence, the scene demonstrates a kind of ritual by which Dawn begins to sever her ties with standard society. She is no longer a virgin of conformity to general or genital expectations. The ring toss scene is an admission that nature encompasses and dwarfs paltry human symbols. Yet Dawn continues to entertain the idea of accepting responsibility for her involuntary actions. She thinks about turning herself in to the police but first wants to learn more about her condition, and so she studies myths and tries to figure out what science may reveal about her mutation. She goes to her school textbook on human anatomy, which has been censored, the diagram of the vagina being covered with a golden star. Delicately, carefully, Dawn reveals what is hidden, what is denied and repressed by her culture, taking another step toward understanding the nature of female bodies and sexuality.

Dawn's connection to the natural world adds to a sense of her ominous power, as one scene showing her tardy arrival for a class hints. Dawn enters the classroom after a teacher has been discussing the evolution of rattlesnakes. The teacher wants to make the subject relevant to her students and says that evolution is about "why you are the way you are." The teacher, in an effort to chide Dawn for her tardiness, says directly to her, "Dawn, this is about you, and you missed it." Lichtenstein doesn't want us to just glean that Dawn is a new product of evolution; he suggests that she has a power akin to a rattlesnake's, and it is no coincidence that Dawn's on-screen appearance during sexually charged scenes is often accompanied by the sounds of a tribal rattle.

In myths about the vagina dentata, women are frequently identified with raw, untamed, wild nature that has yet to be civilized. According to Barbara Creed, a Native story from New Mexico tells of the "house of vaginas" where four "vagina girls" live. Lured by intercourse, unsuspecting men would visit the house. The girls' father, Kicking Monster, would kick men into the house to be devoured. A boy, Killer-of-Enemies, eluded the father and gave the girls a special sour berry medicine that destroyed their vaginal teeth (Creed 106). This Native story places emphasis on how a male hero must defeat dangerous women and subordinate them to masculine authority. The story implies that such subordination is necessary for reproduction, but it also emphasizes that male authority is a hallmark of proper gender relations. Because Dawn in *Teeth* is not finally dominated by a male, she resists this notion of subordination to any male's will. Dawn herself stands for independent feminist authority that may stand outside typical American social customs. This notion of feminist independence is a major trait of contemporary vagina dentata motifs, of which *Teeth* is a striking example.[2]

The Navajo, who seem to be fairly open-minded about the reality of children's sexual play, use the legend of a vagina dentata to prevent boy/girl play from crossing certain lines: "Small boys are sometimes told that the girl's vagina will bite off or injure the penis" (Leighton and Kluckhohn). Most likely, Lichtenstein was aware of such Native practices because in *Teeth*, the heroine Dawn often appears on the screen accompanied not just by rattles but also by the sound of tribal drums. Furthermore, the early dramatic scene in which young Brad is bitten by a much younger Dawn may be specifically indebted to this Navajo legend.[3]

Many old vagina dentata stories emphasize a male perspective that is frequently sexist. The name of the Melanesian goddess Le-hev-hev" Barbara Creed notes, means "That which draws us to It so that It may devour us" (Creed 105). This goddess's name emphasizes that men feel weakened by their desire for women. Barbara Creed also states that in some stories "Chinese patriarchs [believed] that a woman's genitals, apart from offering pleasure," were "executioners of men" (105). Because of male insecurity toward women, the great phallic man was imagined to be a conqueror of the most dangerous female—the one who possessed a vagina dentata. These stories were apparently meant to boost the male ego. Barbara Creed also emphasizes that classical male hero stories about dangerous treks into the underworld contain an erotic dimension. For Creed, Herakles' "taming of the toothy hellhound Cerberus" signifies the "breaking of the vaginal teeth by the hero, accomplished in the dark and hidden depths of the vagina" (106). Herakles is one of the most phallic of the Greek heroes, who overcomes the odds by satisfying the tasks set before him by Hera, who dislikes him because he is the product of her husband Zeus's infidelity, his phallic ubiquity. In a sense, Herakles also tames a dangerous goddess, Hera, who wants to figuratively castrate him and thereby indirectly attack Zeus.

Notions of feminine purity and the need for true love can also be combined with vagina dentata imagery. Barbara Creed observes that a "visual motif associated with the vagina dentata is . . . the barred and dangerous entrance. . . . The suitors who wish to win Briar Rose [Sleeping Beauty] must first penetrate the hedge of thorns that bars their way. Only the prince who inspires true love is able to pass through unharmed" (107). Sleeping Beauty conforms to notions of feminine passivity while she is saved by a noble male suitor. In this story, the vagina dentata becomes strictly metaphorical, but it gives us the feeling that a male has succeeded against a dangerous natural force that may have torn him to shreds. But Sleeping Beauty is primarily a prize and not herself the source of the dangerous dentata; it is a witch who wants to prevent a young woman from fulfilling her reproductive potential and capacity for domestic and sexual pleasure. This witch signifies a cockblocker side of feminine nature, one that rejects masculinity and patriarchal society and that can castrate men if they do not possess enough virile power. In this story, the prince does not have to worry about resentment from his bride once he has achieved success. These old stories often show women as destroyers, enemies of civilized order, and thus figures to be conquered.

In *Teeth*, Dawn becomes a version of an earth goddess through her powerful mutation. She is associated with natural processes, growing awareness, and survival: she is a new beginning as her name suggests. Before Dawn completely accepts her new identity, she fears that she might be a monster, and she seeks to legitimate and understand herself through science, the other major voice of authority in her life that she has been conditioned to accept. Dawn is quite unlucky to end up with the gynaecologist Dr. Godfrey, a supposedly altruistic medical practitioner who turns out to symbolize male corruption that is even more shocking than Tobey's hypocritical Christianity. The new church of the secular world—science—also symbolizes masculine abuse of power, and the gynaecologist's name reinforces that he may like to play God. Dr. Godfrey implies that he is an open-minded and supportive practitioner, a good patriarch who will not judge Dawn's sexual behaviour. But we know that he is up to no good as he removes a rubber glove and starts probing Dawn's vagina with four fingers. He instructs Dawn to relax and receives a sick thrill as he croons: "My goodness, you're tight." The doctor finger-rapes Dawn, ignoring her pleas to stop (Figure 6.1). But Dawn represents a feminine side of nature that will not be ignored. We instantly wonder how many other young women this doctor has violated, and we hope that his punishment will be severe. Without any conscious effort, Dawn's natural reaction occurs. Her vagina dentata springs, and we hear the crunch of fingers being severed from Dr. Godfrey's hand. Dawn even seems to be upset by her natural reflex. This loose-limbed young woman sprawled on the floor trailing the M.D.'s bloody fingers has the audience's sympathy. She is neither wicked nor monstrous; her body is just doing its job. Dawn has embarked on reversing the archetypal vagina dentata stories. She has begun her own hero quest to subdue male exploiters.

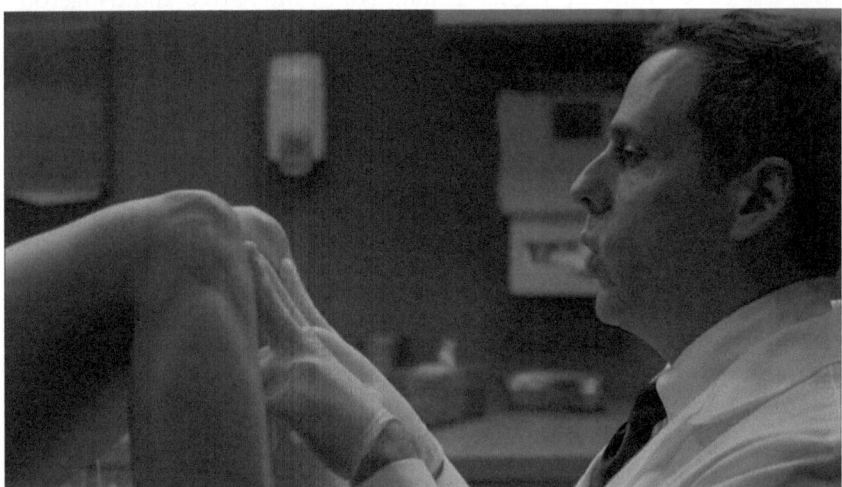

Figure 6.1   Dawn (Jess Weixler), already a victim of environmental toxins, is subjected to abhorrent sexual advances from her doctor, ultimately exacting revenge in *Teeth*.

Another of Dawn's male admirers, a fellow student named Ryan, speculates that Dawn is actually a hyper-sexualized young woman who has not really discovered her true nature. He believes that he will be able to take her on an exploratory journey into the depths of passionate womanhood, and he tries to initiate her the first chance he has. Ryan, somewhat of a Tobey lookalike, tries to seem kind, gentle, and sincere. Having gone through two traumatic experiences with men, Dawn admires Ryan's apparent respect and desire for her. The seduction scene brings up issues of informed consent because he gives Dawn what is most likely a valium and some champagne, but she also appears to long for a non-traumatic sexual experience. Listening to Dawn's somewhat hysterical account of how she possesses a violent vagina, Ryan characterizes himself as a masculine hero who can thwart the dangerous unknown of female sexuality. He comically vows that he will conquer the teeth of her vagina, positioning himself as a Herakles figure. After relaxing Dawn with intoxicants and his little vibrator (a strange item for this young man to possess), Ryan successfully has coitus with Dawn. But the next morning, Ryan reveals his true colours. While Dawn rides on top of Ryan, who foolishly answers a call from a male friend to brag that he is at that moment having sex with her, Dawn then learns that he has wagered that he could seduce her—that he could prove that her vow of abstinence was bogus. When Dawn expresses shock verbally, Ryan's hubris becomes ridiculous as he responds to her, "Your mouth is saying one thing, babe, but your sweet pussy is saying something very different." He is questioning Dawn's knowledge of her own desire and body, saying that he knows her feminine instincts better than she does, a mistake that costs him his penis. Angered by his dishonesty, she accidentally bites off his penis with her vagina, and then expressing regret, she says, "Ah, shit." She is sorry about it, feeling that she has perhaps gone too far. But Ryan's situation quickly degenerates into comic abjection, as we see him quivering with disbelief, ejaculating blood from a tiny stump, and calling for his mother. He does eventually have his penis reattached, but before this can happen he has to be belittled by a doctor who says that the penis is so small it hardly seems to be worth reattaching. The young man with the little penis is weak and insignificant, as is the gynaecologist, in the scheme of things, whereas Dawn exhibits great power.

It isn't until the dephallusation of Brad, the stepbrother, that Dawn exercises her power consciously. She seduces Brad because she is outraged about his treatment of their parents. Not only leaving Dawn's mother for dead, Brad sets his Rottweiler on his father, Dawn's kind-hearted stepfather, Bill, giving him a nasty neck wound. Bill, the only decent male in the film with any depth, does reveal that men in rare cases in the world of *Teeth* may actually be capable of genuine supportiveness and love. To gain vengeance, Dawn briefly complies with Brad's fantasies about her. Brad's despicable character, played by John Hensley, is not unlike the actor's role in the film *Shutter* (2008), in which he plays a macho corporate photographer, an

American on assignment in Japan who participates in the gang rape of a young woman. Near the end of the *Teeth* before Dawn bites off Brad's penis, which is then devoured by his Rottweiler, Mother, we see the nuclear plant spewing extra smoke into the atmosphere. This backdrop of pollution correlates with Brad's demented actions and desires, for apparently he has tried for years to get into his stepsister's pants. The shot after Dawn has bitten off Brad's penis is a brilliant moment in film history, putting Lichtenstein on a par with figures like David Cronenberg and Ely Roth. The film provides a low shot from near the floor beneath Dawn's legs, revealing Brad looking confused, pained, and desperate, clutching at his damaged groin, as she spits out his penis from her vagina. It is a sign of disrespect and female power that is accentuated by the fact that the dog, Mother, then consumes most of the penis, only leaving the tip that has been pierced and studded. This is payback for Brad's vile behaviour toward Dawn's mother; it is vengeance from beyond the grave, carried out by the living daughter.

While the film has elements of comic horror, it also flirts with science fiction because Dawn's mutation apparently results from her exposure to radiation. The film further contains a strong connection to fantasy because Dawn never seems to get bloody. Blood squirts from males, but we are led to believe that not a drop ever touches her. Her amazing ability to avoid the taint of male blood gives the impression that she would never receive any sexually transmitted diseases because of her vaginal power. Perhaps her vaginal mouth is well sealed and spits out any foul matter.

Dawn owns no car but carpools with others as a passenger, or she rides her bike wherever she goes. Dawn, in other words, does not pollute; she stands in opposition to phallic smoke stacks or exhaust pipes. She is a feminist earth traveller at the end of the film, a She-Hulk of sorts who is on the run and carries out acts of poetic justice. She is a person trying to find herself because she does not fit in with society and its masculine and corrupt abuses of womanhood and nature.

In the film's closing scene, Dawn as a hitchhiker tries to exit the automobile of a dirty old man, who symbolizes the antiquated cultural roots of male corruption. This driver of a gas-guzzler Cadillac thinks that he can take advantage of this young woman, and he tries to give the idea that he is sexy with his snake-like tongue. Will Dawn bite off this man's tongue, his fingers, his nose, or his penis? It will all be repulsive and off camera. It will be a new level of male abjection that Lichtenstein has spared us, but we are certain that justified mutilation will ensue. This dirty old man never talks; the final scene is purely gestural. When we know that Dawn will punish this old man, the scene shifts to slow motion, and the heavy beat song, "You Did," reminiscent of cheap pornography, rises. Because the characters exchange no words, the film's conclusion implies that Dawn's body, nature's reaction to masculinized pollution and corruption, speaks for itself.

While *Teeth*'s portrayal of a vagina dentata is based in fantasy and reveals a great deal about cultural attitudes toward gender and the environment, there is the actual possibility that a female could possess a vagina dentata. There is a rare kind of growth, according to Dr. Dean Edell, called a "dermold cyst" that humans can develop; it can cause teeth, bone, or hair to form anyplace on a person's body. It is possible for a person to grow teeth in an armpit or on a breast. It is even possible for a man to grow teeth on his penis. Then why focus on the vagina? That cultures (our own included) do focus on the vagina dentata shows a fixation with women's reproductive organs. The fact that babies dwell for so long within their mothers probably adds to the element of mystery that stimulates people's imaginations on the subject of vaginas and uteruses; here we must note a mythical connection between Mother Earth images and how plants spring from womb-like soil.

Part of the promotion for the film *Teeth* in Britain pushed an element of realism that builds on the idea of dermold cysts. A viral trailer called *Understanding Vagina Dentata* was circulated on the web to help generate interest in the subject. The trailer appears to be a short documentary that explains the biological basis for the condition, and it gives a few first-hand accounts of people who have it. This trailer was entirely a hoax that was created by a marketing team whose goal was to fascinate viewers and perhaps even confuse them. This viral trailer contributed to an urban legend that was widely circulated, and like other legends, it blurred fiction with exaggerated facts.

The people who came up with the trailer *Understanding Vagina Dentata* may have taken some marketing cues from a very successful corporate empire—Starbucks. One Starbucks logo is a modified version of a fifteenth-century image of a Siren, who attracted sailors with her beauty, but then she destroyed them. The Siren has a double tailfin that splits at the crotch and emphasizes the vaginal region. Even though the Siren does not explicitly have vaginal teeth, the image does associate femininity with sexual danger, apparently a theme that generates interest and does not seem to hurt revenue. Starbucks implies that their coffee has the appeal of a mermaid (and consumers can be like lustful sailors), that the coffee comes from trade over the dangerous seas, and that all this attractive danger can be enjoyed in a simple product. Who could resist?

NOTES

1. Jess Weixler's performance as Dawn earned the 2007 Sundance Special Dramatic Jury Prize for Acting.
2. A contemporary term like "man-eater" is not usually used to demonstrate affection. But we also should acknowledge that today the idea of the vagina dentata can carry important feminist connotations; man-eater or castrating bitch may even garner a level of respect. Michael Copperman, writing for *Salon.com*, points out that one of America's most famous people was insulted for being a strong, articulate, and opinionated woman; Hillary Clinton was called "a castrating bitch . . . [because] she insisted on arguing with men as an equal." The notion of gender equality should be a given, but if Hillary Clinton is insulted for making such an assumption, then we have to conclude

that gynophobia is still alive and well; at least, this is the stand that Copper-man takes. But when Hillary Clinton is seen as a "castrator," she is also being accorded respect. Maybe, some men should fear that she could dismantle male privilege or male sexism, and that's why some sexists fear her. If regarded in this manner, terms that relate to the vagina dentata can be empowering.

Similarly, Cjak has designed and screen printed an award-winning vagina dentata shirt that has a strong feminist vigilante message. The shirt sports a large vagina that contains two large and sharp teeth that are dripping blood. Surrounding the vagina, we see the phrase: "THIS TIME *YOU* BLEED." The phrase seems to imply that women are tired of the pain of menstruation, and now the burden will be passed on to men. How this could possibly happen is not explained. But most likely, Cjak wants to state that women have suffered long enough from sexism in American culture, and if men do not watch out, they will be castrated. The shirt implies that men need to be cautious around women; that is, they need to treat women with respect and dignity or else there will be severe consequences.

3. A vibrant recent image called *Vagina Dentata* created by Hundredeyes was directly inspired by the film *Teeth*. In this piece of pop art two legs are opened and fly through the air, and in the place of a vagina we see a toothy mouth that emits blood. Oddly enough, behind the mouth, there is an upside down crucifix at the top of which is a halo of clouds. The whole image has a South-western Native motif because of the colours, the two upright cacti at the bot-tom, the two laughing ravens, and the emphasis on symmetrical design. Does this image mean that the vagina dentata has overpowered Christianity? The artist may have meant this message because in *Teeth* the protagonist Dawn moves beyond the Christian abstinence society to which she belongs. If the vagina dentata has trumped Christianity, does Hundredeyes' image mean that we should now worship powerful women? Could this be the basis for a new form of tribalism that counteracts some Christian misogyny?

WORKS CITED

Cjak. "Vagina dentata shirt." Flikr. 10 May 2009. www.flickr.com/photos/36110578@N07/3351897521/ Web.

Copperman, Michael. "Vagina Dentata or the Story of Sarah the Barracuda." *Salon. com* 5 September 2008. Dream Act—Texas. 10 May 2009. http://dreamacttexas.blogspot.com/2008/09/vagina-dentata.html. Web.

Creed, Barbara. *The Monstrous-Feminine: Film, Feminism, Psychoanalysis.* New York: Routledge, 1993. Print.

Edell, Dean. "Vagina Dentata—In Fact and Folklore." Health Central. 11 May 2009. www.healthcentral.com/drdean/408/13198.html. Web.

Hawthorne, Nathaniel. "The Birth-mark." 1843. *The Heath Anthology of American Literature.* Ed. Paul Lauter. New York: Houghton Mifflin, 2004, 977–987. Print.

Hundredeyes. "Vagina Dentata." Flikr. 11 May 2009. www.flickr.com/search/?s=int&q=vagina+dentata+&m=text. Web.

Kolodny, Annette. *The Lay of the Land: Metaphor as Experience and History in American Life and Letters.* Chapel Hill: University of North Carolina Press, 1975. Print.

Leighton, D. and C. Kluckhohn. *Children of the People: Growing Up Sexually: The Navaho Individual and His Development.* Cambridge, MA: Harvard University Press, 1948. Print.

Psycho. Dir. Alfred Hitchcock. Paramount, 1960. Film.

Rubin, Gayle. "The Traffic in Women: Notes on the 'Political Economy' of Sex." *Toward an Anthropology of Women*. Ed. Rayna R. Reiter. New York: Monthly Review Press, 1975. 157–185. Print.

Shutter. Dir. Masayuki Ochiai. Perf. Joshua Jackson, Rachael Taylor, and Meguni Okina. 20th Century Fox, 2008. Film.

Teeth. Dir. Mitchell Lichtenstein. Lionsgate, 2007. Film.

Understanding Vagina Dentata. YouTube.com. 2008. 13 May 2009. www.youtube.com/watch?v=MgHN1GvF40I. Web. July 2014.

7 The Spirits of Globalization
Masochistic Ecologies in Fabrice Du Welz's *Vinyan*

Georgiana Banita

The cinematic representation of ecological trauma poses questions that are as central to today's green cultural politics as they are resistant to definitive answers. One possible explanation for this conceptual impasse has been the lack of a consistent vocabulary to shore up discussions of eco-trauma that would juxtapose a complex understanding of ecology with the workings of trauma itself. This chapter argues that recent cinematic representations of eco-trauma, particularly with reference to so-called na-tech (natural *and* technological) disasters, shed light on key components of ecological thinking as well as on post-traumatic psychology and aesthetics by illuminating ways in which ecology and trauma are more closely interlinked and interdependent than has so far seemed apparent. Belgian director Fabrice Du Welz's "survival horror" *Vinyan* (2008), a narrative inspired by the Indian Ocean Tsunami of 2004, is my case in point. Not only does the film bring into play notions of mental sustainability that are not commonly associated with ecological thinking; it also introduces a set of useful parameters that help define the confluence of environmental concerns on the one hand and post-traumatic aesthetics on the other. *Vinyan* shows how contemporary ecological thought might be seen as symbiotically intertwined with the capital and cultural flows of globalization, just as the traumatic aftermath of ecological failure is inextricably entangled with a kind of masochist aesthetics whose thematic and structural deployment is one of the film's greatest strengths.

The tsunami that hit South East Asia on December 26, 2004, inflicted widespread psychological trauma,[1] with health workers reporting that virtually everyone in the affected communities incurred some degree of traumatization. The aftermath of the tsunami in Sri Lanka and Indonesia drew particular media attention as well as the support of the World Health Organization, whereas other territories, such as Burma, were largely ignored and their losses underestimated. In an attempt to redress this oversight, *Vinyan* focuses on the effects of the tsunami in the border region between Thailand and Burma.[2] It follows a couple, Jeanne and Paul Bellmer (Emmanuelle Béart and Rufus Sewell), on a journey to recover their son, Joshua—missing and declared dead in the tsunami six months before—after spotting a similar-looking boy in a film about orphans living in the jungles of Burma. Deluded by optimism, the parents (the mother more fervently than the father) believe the boy is held captive by illegal human traffickers. It is, however, more likely that Joshua was killed in the tsunami, a fact that the film never directly acknowledges or describes yet constantly re-enacts in the minds of the parents. Jeanne in particular is plagued by nightmares and by an unhealthy preoccupation with children. Driven by their obsessive and shockingly unrealistic hope that Joshua might still be alive, the couple undertake a dangerous quest to find the boy, in the course of which their trajectory imperceptibly changes: from an external to an interior quest, from a linear, teleological narrative to a muddled visual fantasy that immerses the viewer into the characters' mental landscape (chillingly captured in Benoît Debie's cinematography[3]) where their deep-seated trauma resides.

This descent unfolds in several stages and is emphatically anticipated by the opening credits. Unclear images of dark water and oxygen bubbles rising to its surface, through tangled black streaks of what looks like human hair, are set against a soundtrack that builds to a deafening crescendo of destruction. In the following shot Jeanne emerges explosively from the water, as if trying to escape the remnants of the death it contains (Figure 7.1), yet throughout the film she will be pulled back in repeatedly and entrapped in a symbolic net that includes water-related imagery such as shores, rain, and mud. The opening sequence thus establishes the crucial opposition between the ominous flood and the protagonists' resolve to defy its power.

In due course, however, Paul and Jeanne become estranged from their guide, Thaksin Gao—a somewhat ridiculous Mr. Kurtz bent on extorting as much money from the bereaved couple as possible—and they begin to roam the tropical jungle and deserted villages by themselves. At this point two things become clear: first, that the couple is doomed; and second, that what

Figure 7.1 At the beginning of *Vinyan*, Jeanne (Emmanuelle Beart), a grieving mother, appears resigned to her loss. Yet the initially soothing natural backdrop degenerates very quickly.

is most menacing to them is not the Burmese wilderness, however unfriendly and bizarre its landscapes and inhabitants, but Joshua himself. The boy's indelible image triggers Jeanne's sexual frigidity as well as her sudden resentment of her husband, whom she now holds responsible for Joshua's death—or at least this is the most plausible explanation for her increasingly erratic behaviour. From this moment on, the camera frequently performs gestures of spatial immersion and penetration (through semi-submersion in water, aerial tracking, and sudden dips from above, seamlessly interwoven with ground footage) which signal the protagonists' descent into madness and prompt a viscerally affective response from the viewer. The final segment of the film sees Paul and Jeanne emerge from the subterranean vaults of a temple in ruins, only to be surrounded by an army of feral children who have been watching the couple for days. From above, the at once inquisitive and detached camera records Paul's evisceration at the hands of the children and Jeanne's physical surrender to their enraptured caresses.

Like Aditya Assarat's ironically titled *Wonderful Town* (2007), whose brutal ending mirrors the tsunami's deadly and absurd irruption into everyday life, *Vinyan* traces the contours of a post-apocalyptic landscape with a technique as cataclysmic as the storm itself, deftly capturing the shattered mood of its setting. In both films relationships are savagely severed, bringing the viewer into communion with the people whose lives keep circling back to the locus of trauma. Significantly, unlike the children who are reunited with their parents after being separated by the crashing waves in the more sentimentally conventional film *The Impossible* (2012), Joshua does not resurface. By staging the "return" of children, but not of the one child the parents seek, *Vinyan* has more in common with child-loss films such as *Long Weekend* (1978), *Dead Calm* (1989), and *Antichrist* (2009), in which children or child imagery haunt couples alienated from society and each other as they succumb to the violence of hostile environments.

Building on the link of innocence, brutality, and nature, this chapter analyzes *Vinyan* in the context of trauma in general and of its masochistic after-effects in particular, by way of concepts gleaned from discussions of ecology that exceed the (localized) realm of environmental catastrophe, especially in the aftermath of the transnational turn in eco-criticism.[4] On the face of it, the conjunction of eco-trauma, masochism, and globalization might seem incongruous. My argument is that *Vinyan* positions itself self-consciously so as to reflect not only the global framework of ecological catastrophe but also the ways in which trauma is framed and reduplicated through an elevation of aesthetic form over ethical content. Organizing the film's traumatic vocabularies is a masochistic dynamic whose formative determinant is not, in this case, pre-Oedipal, but an apocalyptic trauma rooted both in ecological disaster (the Indian Ocean Tsunami) and in an ontology of self-harm embodied by Jeanne's self-flagellating quest for the phantom of her son/her marriage/her mental balance. The film rotates its traumatic narrative around a masochistic axis to decentre familiar tropes of globalization and deploys these tropes to produce a discomfiting ecology on an environmental, aesthetic, and psychosexual level. From this angle, the mixture of genre registers in *Vinyan* enacts a sense of fictitious order—reminiscent of masochism's perverted staging of desire—which invites the viewer to re-imagine the links between desire and dislocation, trauma and transnationalism. By focusing on Jeanne's relentless search for her son, we will see how the film brings into play troublesome questions about the relationship between eco-trauma, film, and masochism (with particular attention to its female variant). In a masochistic reversal, while appearing to act on the ideal of maternal self-lessness, the mother performs a sacrifice that restores a fictitious power and control over the real. The protagonists are clearly in a quest for their son, yet they are also reaching out for something less tangible, namely, the existential validation of experience achieved far from their Western origins and from the affluence of upper-middle-class privilege, thereby hoping to restore a measure of dignity. The fantasy of omnipotence the film recounts is thus both self-annihilating and ego-enhancing, countering not only the loss of a child and of an ecosystem, but also a range of geographical dislocations and separations from places and "states" perceived as irrecoverable.

ECOLOGIES: GLOBAL, LOCAL, PSYCHOSEXUAL

In both conception and execution, *Vinyan* is a film shaped by the transnational turn in cinema and, for all its obliqueness, clearly participates in this development.[5] Not only do the film's multivalent aspects formally mimic a globalized aesthetics, but its hybrid crosscurrents (in terms of genre and cinematic tradition) slide around national referents of every kind. *Vinyan* fails to attach itself to any specific national imaginary: At times the film feels decidedly European, then it surprises the viewer with ponderous camera-work reminiscent of contemporary Thai cinema and the culturally hybrid cinematography of Christopher Doyle. This dynamic interrogates not only national identity, on- and off-screen, but also the ecological principles usually attached to local conditions. The film suggests that beyond the locally visible there is another "world" to be apprehended globally, because the spirits who inhabit it, or what Taksin Gao calls "Vinyan," circulate easily below the threshold of embodiment. *Vinyan* gives form to this formlessness, to the larger "virtual" reality spawned by the devastating tsunami, a reality of which the immediately (local) present is only a fragment.

If Jeanne's search for her son is fetishistic, equally fetishistic is the film's deconstruction of the signifying patterns of national trauma. Surely the location of exile provides an external correlative to the couple's condition of psychological and geographical dislocation; yet for all its emphasis on mobility, the film conveys a claustral, enclosed feeling through its spatially oxymoronic structure. Ideas of confinement (in the labyrinth of the Burmese jungle) and escape (from a Western world the Bellmers feel no compulsion to return to) become interwoven. Yet even as far away as Burma they are only a step away from their roots. In her transnational confusion, Jeanne even becomes bound to her son through the register of globalization. The Manchester United T-shirt that Joshua is seen wearing in the Burmese video stands for a kind of youth culture that no society seems impermeable to, not even the remote Burma, a youth culture that has thus become a symbol of globalization itself.[6] And the international parallels do not stop here: Not only does the plight of the wealthy, guilt-ridden parents searching for their child in foreign countries recall the international media exposure of the McCanns after the high-profile disappearance of their daughter, Madeleine; the red Manchester United T-Shirt Joshua is wearing is also the same kit worn by victims in a high-profile UK murder case, that of Holly Wells and Jessica Chapman in 2002.

It is indeed unsurprising that Du Welz should subtly allude to an international setting for his plot in a film haunted by the effects of the Indian Ocean Tsunami of 2004. This catastrophic event, to a greater extent than similar disasters such as Katrina, has invited a plethora of "world-system" approaches that foreground the damage to large-scale environments—both before and after the flood[7]—as well as post-catastrophe international relief (Letukas and Barnshaw 2008). The tsunami triggered a humanitarian crisis on the international stage, provoking an outpouring of generosity from Western countries. The aid provided, however, was often based on inadequate consultation and tended to reinforce the injustices that had contributed to the disaster in the first place (Clark 2007; Korf 2007). Environmental degradation was further worsened by the influx of aid workers—also termed "the second tsunami" (Kennedy et al. 2008)—attempting to implement a hasty rebuilding programme inadequately planned and recklessly wasteful of natural resources.

The specific ecology mobilized in the film is best understood in the terms proposed by Félix Guattari, who extends definitions of ecology beyond environmental concerns to include human subjectivity and social relations. While the representation of the tsunami feeds upon and reinforces public anxieties about ecological disaster, the tense rapport between Paul and Jeanne Bellmer may be read as a warning against the erosion of "personal ecology" in the sense suggested by Guattari, who writes that "it is quite wrong to make a distinction between action on the psyche, the socius, and the environment" (Guattari 41). Instead he proposes an ethical articulation of ecology that comprises three registers: the environment, social relations, and human subjectivity. *Vinyan* can be interpreted as an illustration of Guattari's "social ecosophy"—which consists in "developing specific practices that will modify and reinvent the ways in which we live as couples or in the family" (34)—and of his "mental ecosophy," leading us "to reinvent the relation of the subject to the body, to phantasm, to the passage of time, to the 'mysteries' of life and death" (35).[8]

This extended ecology is just what we find acted out, if on a more grotesque level, in *Vinyan*. The film delivers a compelling study of trauma not
only as the effect of a world-shattering event (public or personal), but especially as a key to rethinking the dynamics of emotional family ties, such
as conjugal relations and motherhood. The trauma suffered by Paul and
Jeanne Bellmer coincides not only with the shock caused by their son's disappearance in the tsunami, but above all with a complex process of ecological enlightenment. They relinquish their sheltered Western lives and later
the (relative) safety of the Thai metropolis to plunge into the perils of the
Burmese jungle. They begin by searching for a boy dressed in a red Manchester United jersey only to gradually learn that in nature the distinctions
among children (and other creatures) are much less stable than they may
have assumed, just as—according to their guide Taksin Gao—all the people
who died a violent death during the tsunami become the same vengeful
spirit, "Vinyan."

This transition from extreme individualism toward "deep ecology"
becomes most palpable in Jeanne's evolution from being the mother of a single (irreplaceable) son to serving as totemic muse to the dozens of orphans
she encounters at various stages in her quest: playing by the side of the road;
throwing stones at bodies washed ashore by the flood; eerily watching her
from the spectacularly backlit docks along Burma's shores; following the
couple from beneath the verdant camouflage of the jungle. The narrative arc
of the film transforms Paul and Jeanne from a couple convinced of the indi-
viduality and irreplaceability of their only child into two constantly bicker-
ing people who no longer relate to each other, lose control of their actions,
and react only to the hostilities of nature. Yet Jeanne acquires a responsive-
ness to her surroundings that sets her apart from her husband and allows
her to be easily inducted into the savage group she encounters and to readily
accept, in a way that betrays her self-flagellating inclinations, the surreal or
downright hostile ecosystem of the post-catastrophic Burmese jungle.[9]

THE CONCUSSED CAMERA OF MASOCHISM

Vinyan, I will argue in this section, is implicated within what it describes
thematically in that it replicates eco-trauma at a structural level.[10] Both the
film's characters and its settings are enmeshed in the economy of trauma.
Vinyan conveys the trauma of mass casualty implied in the opening credits
and in the locals' commemoration rituals along the shores by inflicting on
the spectator the trauma of witnessing murder. In this sense, the final scene
functions as what Joshua Hirsch would term "traumatic afterimage," which
reproduces the shock of the original atrocity (Hirsch 19). Even well before
the film's finale viewers easily identify motifs and themes that have been
abundantly deployed in trauma cinema, above all the crossing of the realist

mode on the one hand with fantastic, oneiric traditions on the other, doubling and refracting each other.[11] The film is also replete with visual motifs
and images saturated with still other images: not only visual artefacts (such
as the Burma video and Jeanne's Joshua-themed photo album) but also
a series of scenes originating in what Bruce Kawin termed "mindscreen"
(Kawin 1978): Jeanne encountering Joshua in a room drenched in red rain,
with Paul glimpsing the boy in the middle of the dark jungle.

The fantastic elements that saturate *Vinyan*, in conjunction with the
sharp, disorienting montage, match the structure of trauma as a disruption of temporality (Caruth, *Trauma* 4–5; Whitehead 6). According to Janet
Walker, "trauma cinema" deals with world-shattering public or personal
events "in a nonrealist mode characterized by disturbance and fragmenta-
tion of the film's narrative and stylistic regimes" by "representing reality
obliquely, by looking to mental processes for inspiration, and by incorporating self-reflexive devices" (Walker 19). Following this model, *Vinyan*
departs from a linear narration of events, confusing not only time spans but
also the psychological axes around which the protagonists' minds rotate,
wildly mixing motivations and affections like colours on a palette, with
impatience and a short-sightedness that testifies to the couple's lack of
direction. Precisely what the film contributes to these by now conventional
trauma aesthetics I will contend with later, but for the moment I want to
situate the film in a broader context of influences and genre categories, as
their mixture will ultimately define *Vinyan*'s innovative take on eco-trauma.

"As trauma is less a particular experiential content than a form of experience," Joshua Hirsch notes, "so posttraumatic cinema is defined less by a
particular image content . . . than by the attempt to discover a form for presenting that content that mimics some aspects of posttraumatic consciousness itself, the attempt to formally reproduce for the spectator an experience
of suddenly seeing the unthinkable" (Hirsch 19). And insofar as what is
visually thinkable is partly constituted by the conventions of film genre,
Vinyan attempts to upset these genre expectations. As in his debut feature
Calvaire (2004), Du Welz draws upon, abbreviates, and in some cases sub-
verts a number of traditions and influences including the survival narrative,
the horror film, the adventure story, the descent into madness, and the Con-
radian genre initiated in cinema by *Apocalypse Now* (1979). Like Nicolas
Roeg's *Don't Look Now* (1973), with which it shares a blasphemous vision
of the cult of the child, *Vinyan* employs archetypal symbols that "recall the
possibility of individual transcendence, the journey of a hero to the land
of the dead from which he is to return with special knowledge and experience to the living world" (Palmer and Riley 17). In both cases, however,
the protagonists are ultimately trapped in an un-mythic, Gothic world—see
especially the final act's derelict temple—from which there is no escape.

In its terrifying displays of graphic carnage, however, *Vinyan* is pre-
dominantly a horror film. It corresponds to this genre by virtue of being
"expressly repulsive" (Carroll 158) and relishing in its own "disturbing,

distasteful, or even downright unacceptable" transgressions (Gelder 5). Yet it also borders on the territory of a cinema genre that is less frequently associated with the eco-mode, namely, the fantasy film. On the surface, *Vinyan* is a horror film that re-enacts the evolution of the genre from a focus on otherworldly monsters toward human threats (from uncanny images of the missing child as revenant to the orgiastic mud bath of his demented mother), from understated destruction to graphic violence. The more human the threats depicted, the more otherworldly the film becomes, as it segues into a magical-realist mode that inhabits the characters' innermost thoughts only to ultimately evacuate them through Paul's literally gut-wrenching death at the hands of the feral children. The film thus evokes visceral horror, disrupts genre categories, and carves out a mysterious place both on the screen and in the spectatorial relation that consists in equal parts of horror and fantasy. One could argue that *Vinyan* is, strictly speaking, neither a horror nor a fantasy film. Rather, the question I pose is How does the film access discourses of horror and fantasy to confront the representation of ecological trauma—in the broader sense discussed earlier?

In a rich constellation of competing films that take up hysterical post-traumatic themes, *Vinyan* is both original and compelling. Released in 2008, the same year that Michael Winterbottom's *Genova* (2011) followed a family's move to Italy after a bereavement (inviting unflattering comparisons with Roeg's *Don't Look Now*) and one year before Lars von Trier's scandalous *Antichrist*, whose opening montage shockingly juxtaposes a couple's lovemaking with the accidental death of their child, *Vinyan* combines elements of both films. It uses the conventionality of a "contextualized" family drama to enhance the masochistic effect that von Trier's self-mutilating aesthetics seek to achieve in vitro. With its obsessive torturing of a family caught up in a maelstrom of random violence, *Vinyan* recalls Michael Haneke's work in *Funny Games* (1997, 2007) and anticipates Haneke's painfully subtle treatment of guilt (individual, collective, and unattributable) in his 2009 masterpiece *The White Ribbon* (*Das Weiße Band*). Both *The White Ribbon* and *Vinyan* entail the participation of uncanny child figures in the persecution of the protagonists; child figures in general, especially absconding children whose disappearance has the capacity to throw families into disarray, have inspired several productions of note in recent years. *Gone Baby Gone* (2007, dir. Ben Affleck), released on the heels of the McCann case, and *Changeling* (2008, dir. Clint Eastwood) stand out, although both these films' ambiguous endings are less memorable than *Vinyan*'s open-ended final scene of visceral motherhood, whose sensuality recalls the censored cub-bonding shot (of a mother tenderly licking her fresh foetus) in David Cronenberg's *The Brood* (1979).

In the film's final scene, especially in Paul's gruesome death, we catch a glimpse of what Adam Lowenstein calls an "allegorical moment," defined as "a shocking collision of film, spectator, and history where registers of bodily space and historical time are disrupted, confronted, and intertwined"

(Lowenstein 2). Lowenstein identifies the core of the allegorical moment in the process of embodiment, "where film, spectator, and history compete and collaborate to produce forms of knowing not easily described by conventional delineations of bodily space and historical time" (2–3). *Vinyan* crystallizes the allegorical moment as the very opposite of embodiment, namely, as a set of variations on evisceration. The image of evisceration challenges the concepts that govern the study of trauma and its representation: memory, mourning, and working through. The film does not repeat the polarized pattern of acting out versus working through, allowing us to perceive the post-traumatic condition in the correspondences between these seemingly competitive modes. In the end trauma becomes, as Cathy Caruth has written—here even literally—"a voice that cries out from the wound" (Caruth, *Unclaimed* 2).

Through its ambivalent fantasy of mass motherhood and mass murder, the film also reveals a more discrete wound, a split at the heart of trauma, namely, the rift between the desire to embrace it (Jeanne's voluptuous orgy with the child savages attests to this) and a logical resistance to it. As a practitioner of rational thinking, Paul may well be able to accept and overcome his own psychological sequelae, yet he cannot learn to live with Jeanne's as well. His death can be read as an act of retribution for (to put it with Jeanne) "letting Josh go," that is, smoking out his own demons so as to assuage his trauma. The film's relation to traumatic experience remains, however, emphatically anti-redemptive. To borrow some of Adam Lowenstein's critical terms, "rather than offer reassuring displays of artistic 'meaning' . . . in the face of historical trauma," *Vinyan* requires "that we acknowledge how these impulses to make productive meaning from trauma often coincide with wishes to divorce ourselves from any real implication within it" (Lowenstein 8–9). It is this wish to divorce himself from Joshua's memory that Jeanne diagnoses in Paul and later condemns as matrimonial alienation.

Jeanne, by contrast, never lets go of Joshua's memory and of the tsunami's sensorial aftermath contained in images of flood, drowning, hunger, and spatial disorientation. The eco-traumatic diminishment in *Vinyan* is thus both corporeal[12] and cinematic, as traumatic experience "cannot be organized on a linguistic level, and this failure to arrange the memory in words and symbols leaves it to be organized on a somatosensory or iconic level: as somatic sensations, behavioural re-enactments, nightmares, and flashbacks" (van der Kolk and van der Hart 172). While it does not entirely abandon narrative, *Vinyan* stages a collision between narrative and the purely iconic, a collision from which the film derives its shock effect. The plot line disintegrates to the point where narrative and hysteria are indistinguishable, Jeanne providing the valve through which terror flashbacks are inserted into the narrative chronology. Linear time collapses or is experienced as uncontrollable; the past appears too immediate, presenting itself uninvited; nature and culture converge; psyche becomes environment and vice versa.

Feeling that conventional filmic narration is inadequate to represent both the tsunami's apocalyptic force and the couple's doomed quest, Du Welz turns to fantasy as an alternative to realistic storytelling, drawing on the fantasy genre's affinities with the eco-traumatic mode. In the Freudian tradition, fantasy denotes an "imaginary scene in which the subject is a protagonist, representing the fulfilment of a wish (in the last analysis, an unconscious wish) in a manner that is distorted to a greater or lesser extent by the defensive processes" (Laplanche and Pontalis 314). That is to say, it is a non-veridical construction based on a real event, which "is not the origin of the fantasy but rather the grain of sand around which the pearl of fantasy is deposited" (Walker 9). *Vinyan* shows that far from impoverishing fantasy life by generating an endless repetition of mechanical flashbacks, trauma can stimulate it by creating elaborate imaginary structures. Du Welz seems to share the perception of trauma as the breakdown of symbolic resources, narrative, and imagery, yet rather than merely doubting the possibilities of representation, this concept of trauma spurs it to generate new forms of understanding traumatic experience. In this sense, the entire film is a flash-back, a transformative remembrance collecting a series of scenes that pertain to the actual trauma but do not coincide with it. As a common feature of traumatic memory, disremembering is defined by Janet Walker as the process of "conjuring mental images and sounds related to past events but altered in certain respects," or "remembering with a difference" (Walker 17). "Disremembering" connotes, of course, not just impaired memory but also dismemberment. In *Vinyan* such disremembering (what Jeanne repeatedly refers to as "letting go") is not, as in Walker, a "survival strategy" (19), but a murder for which capital punishment is meted out. This brutality can be taken as a consequence of an extreme post-traumatic stress disorder (PTSD): Both Paul and Jeanne, I believe, might be seen as suffering from vicarious trauma induced by viewing the Burmese orphan film showing the aftermath of the tsunami, and Jeanne's idea that Joshua may be among the orphans could have been triggered by this partly ecological trauma.

Significantly, it is Joshua's image in a Manchester United T-shirt, as seen in the film, that becomes lodged in the mother's mind where it remains indefinitely dissociated, returning in pathological forms (such as nightmarish visions). Cathy Caruth associates PTSD with "the often uncontrolled, repetitive appearance of hallucinations and other intrusive phenomena" (Caruth, *Unclaimed* 11), especially flashbacks—a phenomenon that lends itself very well to cinematic adaptation. However, Jeanne does not experience flashbacks but flash-*forwards*—hallucinations and fantasies that are never repetitive but highly aestheticized and controlled. These fantasies run forward rather than backward partly because Joshua's death destabilized the traditional setup of trauma relations in the family. Usually it is the *children* of survivors who recall events that they experienced only indirectly (through a process that Marianne Hirsch has called "postmemory"). In *Vinyan*, this dynamic is reversed, with the parents imagining the child as he *would* be, rather than as he *was*. As a result, the entire film appears as an immersive, almost hallucinatory experience, moving haphazardly around a translucent, illogical axis rather than following a chronological flow of events. The narrative recoils, pointing backward, sideways, distrusting itself, and defying viewers' expectations. "Fantasy" is thus transformed from a mere plot device into the basic mode of narration. The camera itself internalizes eco-traumatic consciousness by taking prolonged, stationary shots of the lush jungle, resting with painful insistence on the rambling movements of the couple through its still landscape. These long takes, Ban Wang has argued, allude covertly to an unassimilable traumatic shock that punches a hole in the apparatus of visual representation: "The camera seems to be staring into an abyss of misery and pain, and seemingly unable to get on with the story, it records every detail and gesture, soaking up every perceivable trace, hue and shape. As if haunted by the dream recurrence of a shocking event, the camera cannot help bearing witness to what has been out there. The long take thus functions as a metaphor for the traumatized patient, who, when asked to tell a coherent story, is repeatedly and helplessly seized by a singular, persistent image" (Wang 238).

I want to propose the dynamic of masochism as a heuristic in describing the correspondence *Vinyan* stages between its technology of representation and its thematic interest in the eco-traumatic consciousness. *Vinyan*, I argue, both depicts masochistic behaviour and creates a fictional textuality around Jeanne's desire, which lends credence to Gilles Deleuze's view that masochism is above all a formal system through which a parallel world is created to accommodate the transgressions of masochistic fantasy (Deleuze 16–17). The fantasy at the core of the film tries to wrest power and agency away from the environment and to restore it to the parents, who are left broken and rudderless by the disappearance of their son. Jeanne recovers this agency by assuming a position of power cemented by acts of violence against her husband and herself. The film itself undergoes a similar transformation in that it starts from a conventional narrative structure which it then gradually takes apart with methods as drastic as Jeanne's in her search through the jungle. Specifically, at the same time that the film reveals Jeanne's compulsive self-victimization it also demonstrates a structural complicity with her behaviour. If this masochistic streak cannot always be spelled out, neither can it be denied, vacillating as it does between the obvious (in the film's more unabashedly violent moments) and the metaphysical. Jeanne's masochistic tendencies surface most clearly in her morbid and reckless decisions to pursue the search for Joshua at the expense of her own mental sanity and physical well-being: even after it becomes clear that the locals cannot tell any white child apart from any other; after Paul adamantly refuses to go on (she secretly pays the guide to continue the search); after their boat departs in the middle of the night, leaving them and the hapless guide to fend for themselves, without food or water, among the savages; finally, even beyond the point where the orphans who roam the jungle make their murderous intentions clear by treating them with playfully callous indifference when approached by the weak and increasingly deranged Jeanne.

At no time, however, does Jeanne let slip her virtuous mask of messianic motherhood. What she interposes as a front for her masochism is a façade of normalcy, of empathetic action, and of fidelity to the family (her own and the family of man). The performance of her suffering can be read not only as resistance to the power of nature but also as an overbearing form of colonial control. "It is important to remember," Gaylyn Studlar cautions, "that the masochistic search is not for pain in and of itself but for a specific position from which the subject can control desire in relation to the object and repair narcissistic wounds by preconditioning (or paying for) pleasure through suffering or humiliation" (Studlar 40). By seeking out pain in every form, Jeanne not only tries to find compensation for her bereavement but also a firmer grip on her troubled transnational entanglements. What the film's perverse scenario unconsciously replays is Jeanne's guilt, part of a larger, subtly implied Western shame, at the inequity of social and *eco*nomic relations separating the Bellmers and their world from the country whose perils they accept with such abandon. The film's ending occasions, in fact, a return to the primal scene of this inequity. Jeanne's steely will is thus imperious and imperial at the same time, for even though she subjects herself to the local horrors she lifts herself above them by choosing the path of greatest resistance.

In other respects as well, Jeanne matches the masochistic profile, whereas Joshua is perfectly suited to play the part of her elusive object of desire (in absentia). We never see the child prior to the tsunami, which reinforces the impression that Jeanne endows the object of her obsession with limited power, so that she can easily direct and control it, a fact that the child's physical absence from the film clearly confirms. But also in typical masochistic fashion, and somewhat paradoxically, Jeanne does bestow enormous power on an object that she patently lacks all connection with: Nowhere do we witness an emotional bond between the two; when he does appear in her memories or hallucinations, Joshua frightens her more than anything else; and even in the grainy Burma that sends the couple on their perilous journey, Joshua is looking and moving away from the camera. Finally, the film mobilizes narrative elements that Gaylyn Studlar associates specifically with masochistic structuring, such as "the 'frozen' doubling of scenes, the intensely repetitive visual and aural components, cyclical narrative structure, and a temporality centered around suspense and anticipation" (Studlar 37).

Even more interesting than the overlap between elements of eco-trauma cinema, as discussed earlier, and the aesthetics of masochism is *Vinyan*'s complex gender dynamics. In its realistic moments the film encourages viewers to perceive the physical world as the husband does, while retaining a quality that eludes definition and that we gradually begin to associate with the wife and mother. The spectrum of the film's genre hybridity also points toward a resistant, unruly form of feminine desire. It does so by reworking

Laura Mulvey's assertion that "sadism demands a story" (Mulvey 14) into "masochism doesn't." That masochism precludes narrative, is further accentuated by strange temporal glitches and sharp contrasts in the way the film promiscuously slides out of one consciousness and into another. Yet despite the sexual connotations in a range of imagery and in the film's seemingly clear-cut gender typology, its particular monstrosity is closely aligned with the collapse of traditional gender roles, with Jeanne emerging as a dominant figure who forces her husband into a submissive position. Her asexuality further reinforces her status as the one half of the couple that merely tolerates the other half. Yet there is more to Jeanne's superficially asexual desire (for her child, for the quest as such, for self-destruction). As Studlar observes, "although female sexuality may be figured differently in masochism's fantastic scenarios (i.e., without depiction of overt 'sexual acts'), it does not mean that eroticized pleasure is not made available or that perverse sexuality is not structurally integrated into them" (Studlar 44). Jeanne's physical encounter with the children in the final scene exposes and redeems her masochistic inclinations (Figure 7.2), proving that Joshua was a sexual simulacrum all along, despite the woman's lack of overt interest in sex for the duration of the search.

A true masochist to the very end, even as she experiences erotic pleasure Jeanne cannot help denying its existence: In the final scene of carnage, she brutally severs her marital bond (all it takes is a look and a strategic positioning in the midst of the lynching mob) in a gesture that can be read as a prolongation of her earlier sexual rebuffs. This is juxtaposed with her expressionless acceptance of the children's attentions, which materialize in a long, silent scene of exfoliating sensuality. Drained of motherly qualities, Jeanne embraces this group physicality and dissolve into nature as the perfect scenario of anticipation and suspension—in classic masochistic fashion—that does not require the object of desire to be present, feeding instead on

Figure 7.2   Herself orphaned by the tsunami, and all her ties to the Western world cut off, Jeanne relishes in the orphans' rapt attention.

voluptuous longing. In fact, at the moment of false identification—that is, the point where another child is proffered as the missing Joshua—the closing gap between subject and object of desire threatens to create an intimacy that both Jeanne and the viewer reject: For both, gratification comes too early, and the viewers realize now, if they haven't already done so, that their own spectatorial desires are as immutably fixed as Jeanne's on the impossible retrieval of a loss that in the film's closing moments finally becomes "naturalized."

CONCLUSION

In its most profound moments this unusual film mines the emotional underpinnings and formal strategies of eco-trauma by drawing on the aesthetic of masochism recast in a global, post-catastrophic light. Jeanne fixates on her lost child partly in a compulsive desire to undo the damage of the tsunami on the world and on her marriage, partly as a response to what she sees as the inassimilable savagery of the Burmese jungles, synecdochic of a world she, the generous European, wishes to annex. Joshua thus remains a fetishized object at the same time as he is projected into a strangely disembodied *enfant sauvage*, a transnational symbol of transgression, of the inassimilable and the foreign, of desire and symbolic displacement. Once the audience has been trapped inside Jeanne's narrowing point of view—marked by encroaching mental illness and psychic disintegration—it becomes clear that what the mother experiences is the fallout of emotions she has kept in check since Joshua's disappearance, emotions that are on the verge of exploding with a forcefulness that recalls the tsunami itself. The storm, then, whose representation in the film is oblique and confined to the shadows of the psyche, comes to signify an insidious form of repression, beyond the individual and the national, which manifests itself in an explosion of self-destructive violence, an image that is tied to the ecological and economic factors that added to the devastation. In this sense, the flood represents both a fetishistic substitute and a simulacrum of the apocalypse.

Significantly, nowhere in the film is the Bellmers' "home" or nationality invoked. The storm has swept away all their cultural ties, so when Paul, in a fit of drunken desperation, entreats the boy to "go home," and Jeanne utters her final accusation to Paul ("you let him go"), one has to wonder what "home" they might be referring to: The only child Jeanne is seen putting to bed in a domestic setting is an orphan who cannot sleep, and who fails to reciprocate her affections. "Going," here, carries the meaning implied by the lanterns that the locals light and propel into the open night sky in a memorial service for the dead: they guide the spirits of the dead into the afterlife. So does Joshua, whose main function in the film is to help his parents find their way to a "home" as the tsunami's aftermath has left them emotionally destitute. And like the other victims who suffered violent, premature deaths, Joshua's spirit becomes angry; he becomes, as the locals put it, "Vinyan." Amid small armies of feral children, far from restoring the comforts of family and domesticity, the Bellmers' journey points to the irreversibility of the tsunami's devastation, which serves not as a momentary failure in the structure of things, but as an opening through which a vulnerability at the heart of human nature is exposed. Cinema, Du Welz suggests, should not linger on the ghost's refusal to "go home"—the topographies of home having been thrown into disarray, literally, by the tsunami—but accept what Homi Bhabha would call the "unhomed" (9) spaces of the eco-traumatic condition and deploy their circular, self-flagellating aesthetics to renegotiate their proper ecology.

NOTES

1. The Indian Ocean Tsunami was triggered by an earthquake measuring 9.0 on the Richter scale, which opened the ocean floor 150 miles off the coast of Sumatra, Indonesia, on December 26, 2004. The tsunami was the largest and most destructive in recorded history, resulting in a quarter of a million casualties in twelve countries (Letukas and Barnshaw 2008).
2. Although *Vinyan* is very much a film "about" the 2004 Indian Ocean Tsunami, no explicit representation of that disaster is featured. This suggests that Welz would agree with Joshua Hirsch who has argued in relation to Holocaust cinema that "all historical representation is . . . limited in at least three ways: by signification (the ontological difference between the reality and the sign, including the memory sign), by documentation (limited documentation of the past), and by discourse (limited framing of documents by the conventions of discourse)" (5). Unlike the Holocaust or 9/11, the Asian tsunami represents a rupture that is mourned outside of culture, as it was both natural and removed. The film seems to be in search of the traumatic wound that led to its creation and triggers its plot.
3. Cinematographer Benoît Debie is to be credited for the hand-held, spinning camera technique of Gaspar Noé's highly experimental features *Irreversible* (2002) and *Enter the Void* (2009); *Vinyan*'s camerawork seems downright conservative by comparison.
4. On new directions in the confluence of ecocriticism and transnationalism, see Heise 2008.
5. For an overview of cinema that addresses issues of globalization understood primarily in economic terms, see Zaniello 2007.
6. Regarding globalization from the perspective of youth culture enriches the dynamics that already define the field. In the words of Raymond Grew, "Discussion of imperialism, world systems, and dependency often proceed from the top down, emphasizing dominance and subordination, the use of power, and the intentions of map-drawing statesmen and profit-seeking entrepreneurs and corporations. Studying children allows the scholar to start from a kind of neutrality that allows a fresh look" (Grew 854).
7. Environmental factors (caused by socio-technical failures) contributed to the scale of the natural disaster (Williams 2008). In the aftermath of the 2004 tsunami, scientists confirmed that "aside from the massive loss of human lives, the other most obvious casualties of the tsunami were marine life and ecosystems" (Atienza 17). Vanished island and coastline erosion reveal the extent of the damage on a larger scale.

8. For a discussion of Guattari's notion of mental ecologies as a heuristic for interpreting other environmentalist narratives, see Banita 2008. '
9. It is this state of indiscriminate abandon on her part that echoes the core tenets of "deep ecology": "Deep ecology is concerned with encouraging an egalitarian attitude on the part of humans not only toward all members of the ecosphere, but even toward all identifiable entities or forms in the ecosphere. Thus, this attitude is intended to extend, for example, to such entities (or forms) as rivers, landscapes, and even species and social systems considered in their own right" (Sessions 270).
10. The concept of trauma has only recently been discussed in relation to film. See Walker, Kaplan and Wang, and Joshua Hirsch (especially his view on the function of post-traumatic film as vicarious trauma).
11. As Christian Metz has stated, "more than other arts, or in a more unique way, the cinema involves us in the imaginary" (Metz 45). Or, as Joshua Hirsch puts it: "Both formally and technologically (through the projection of giant images in the dark), film imitates and engages the experience of processing what has been witnessed through mental imagery, memory, fantasy, and dreams" (Hirsch 7).
12. E. Ann Kaplan and Ban Wang describe prevalent ways of understanding trauma as "engraved on the body, precisely because the original experience was too overwhelming to be processed by the mind. . . . Thus trauma is viewed as a special form of bodily memory" (Kaplan and Wang 5).

WORKS CITED

Antichrist. Dir. Lars von Trier. Criterion Collection, [2009] 2010. Film.
Apocalypse Now. Dir. Francis Ford Coppola. Paramount, [1979] 1999. Film.
Atienza, Yvette. "Tsunami's Impact on the Environment." *Asia Pacific Biotech News* 9.1/2 (2005): 17–20. Print.
Banita, Georgiana. "The Ecology of Love: Reading Annie Dillard with Félix Guattari." In *An Un(Easy) Alliance: Thinking the Environment with Deleuze/Guattari*. Ed. Bernd Herzogenrath. Cambridge: Cambridge Scholars Press, 2008. 297–313. Print.
Bhabha, Homi. *The Location of Culture*. London: Routledge, 2004. Print.
The Brood. Dir. David Cronenberg. Metro Goldwyn Mayer, 1979. Film.
Calvaire. Dir. Fabrice Du Welz. Palm Pictures, 2004. Film.
Carroll, Noel. *The Philosophy of Horror, or: Paradoxes of the Heart*. London: Routledge, 1990. Print.
Caruth, Cathy, ed. *Trauma: Explorations in Memory*. Baltimore: Johns Hopkins University Press, 1995. Print.
Caruth, Cathy. *Unclaimed Experience: Trauma, Narrative, and History*. Baltimore: Johns Hopkins University Press, 1996. Print.
Changeling. Dir. Clint Eastwood. Universal Studios Home Entertainment, 2008. Film.
Clark, Nigel. "Living through the Tsunami: Vulnerability and Generosity on a Volatile Earth." *Geoforum* 38.6 (2007): 1127–39.
Das weiße Band: Eine deutsche Kindergeschichte. Dir. Michael Haneke. Sony, 2009. Film.
Dead Calm. Dir. Phillip Noyce. Warner, 1989. Film.
Deleuze, Gilles. *Masochism: An Interpretation of Coldness and Cruelty*. New York: George Braziller, 1971. Print.
Don't Look Now. Dir. Nicolas Roeg. Paramount, 1973. Film.
Enter the Void. Dir. Gaspar Noé. IFC Films, 2009. Film.

Funny Games. Dir. Michael Haneke. Tartan, 1997. Film.
Funny Games. Dir. Michael Haneke. Warner, 2007. Film.
Gelder, Ken. "Introduction: The Field of Horror." In *The Horror Reader.* Ed. Ken Gelder. London: Routledge, 2000. 1–10. Print.
Genova. Dir. Michael Winterbottom. Ascot Elite, 2011. Film.
Gone Baby Gone. Dir. Ben Affleck. Miramax, 2007. Film.
Grew, Raymond. "On Seeking Global History's Inner Child." *Journal of Social History* 38.4 (2005): 849–58. Print.
Guattari, Félix. *The Three Ecologies.* Trans. Ian Pindar and Paul Sutton. London: Athlone Press, 2000. Print.
Heise, Ursula. *Sense of Place and Sense of Planet: The Environmental Imagination of the Global.* New York: Oxford University Press, 2008. Print.
Hirsch, Joshua. *Afterimage: Film, Trauma, and the Holocaust.* Philadelphia: Temple University Press, 2004. Print.
Hirsch, Marianne. *Family Frames: Photography, Narrative, and Postmemory.* Cambridge: Harvard University Press, 1997. Print.
The Impossible. Dir. J.A. Bayona. Apaches Entertainment, 2012. Film.
Irreversible. Dir. Gaspar Noé. Tartan, 2003. Film.
Kaplan, E. Ann and Ban Wang, Eds. *Trauma and Cinema: Cross-Cultural Explorations.* Hong Kong: Hong Kong University Press, 2004. Film.
Kawin, Bruce. *Mindscreen: Bergman, Godard, and First-Person Film.* Princeton: Princeton University Press, 1978. Print.
Kennedy, Jim, Joseph Ashmore, Elizabeth Babister, and Ilan Kelman. "The Meaning of 'Build Back Better': Evidence from Post-Tsunami Aceh and Sri Lanka." *Journal of Contingencies and Crisis Management* 16.1 (2008): 24–36. Print.
Korf, Benedikt. "Antinomies of Generosity: Moral Geographies and Post-Tsunami Aid in Southeast Asia." *Geoforum* 38.2 (2007): 366–78. Print.
Laplanche, Jean and J.-B. Pontalis. *The Language of Psychoanalysis.* Trans. Donald Nicholson-Smith. New York: Norton, 1974. Print.
Letukas, Lynn and John Barnshaw. "A World-System Approach to Post-Catastrophe International Relief." *Social Forces* 87.2 (2008): 1063–87. Print.
Long Weekend. Dir. Colin Eggleston. Synapse, [1978] 2005. Film.
Lowenstein, Adam. *Shocking Representation: Historical Trauma, National Cinema, and the Modern Horror Film.* New York: Columbia University Press, 2005. Print.
Metz, Christian. *The Imaginary Signifier.* Trans. Celia Britton, Annwyl Williams, Ben Brewster, and Alfred Guzzetti. Bloomington: Indiana University Press, 1982. Print.
Mulvey, Laura. "Visual Pleasure and Narrative Cinema." *Screen* 16.3 (1975): 6–18. Print.
Palmer, James and Michael Riley. "Seeing, Believing, and 'Knowing' in Narrative Film: *Don't Look Now* Revisited." *Literature/Film Quarterly* 23.1 (1995): 14–25. Print.
Sessions, George. *Deep Ecology for the Twenty-First Century: Readings on the Philosophy and Practice of the New Environmentalism.* London: Shambhala, 1995. Print.
Studlar, Gaylyn. "Masochistic Performance and Female Subjectivity in *Letter from an Unknown Woman.*" *Cinema Journal* 33.3 (1994): 35–57. Print.
van der Kolk, Bessel and Onno van der Hart. "Intrusive Past: The Flexibility of Memory and the Engraving of Trauma." In *Trauma: Explorations in Memory.* Ed. Cathy Caruth. Baltimore: Johns Hopkins University Press, 1995. 158–183. Print.
Vinyan. Dir. Fabrice Du Welz. Sony, 2008. Film.
Walker, Janet. *Trauma Cinema: Documenting Incest and the Holocaust.* Berkeley: University of California Press, 2005. Print.

Wang, Ban. "Trauma, Visuality, and History in Chinese Literature and Film." In *Trauma and Cinema: Cross-Cultural Explorations*. Ed. E. Ann Kaplan and Ban Wang. Hong Kong: Hong Kong University Press, 2004. 217–40. Print.

Whitehead, Anne. *Trauma Fiction*. Edinburgh: Edinburgh University Press, 2004. Print.

Williams, Stewart. "Rethinking the Nature of Disaster: From Failed Instruments of Learning to a Post-Social Understanding." *Social Forces* 87.2 (2008): 1115–38. Print.

Wonderful Town. Dir. Aditya Assarat. Kino International, 2007. Film.

Zaniello, Tom. *Cinema of Globalization: A Guide to Films about the New Economic Order*. Ithaca: ILR Press, 2007. Print.

8 Love in the Times of Ecocide
Environmental Trauma and Comic Relief in Andrew Stanton's *WALL-E*

Alexa Weik von Mossner

Global ecocide is usually envisioned as catastrophe. This is why themes of chaos and disaster dominate most feature films that consider large-scale environmental change and a total or near-total destruction of the biosphere. Roland Emmerich's doomsday disaster movie *2012* and John Hillcoat's dystopic film adaptation of Cormac McCarthy's Pulitzer Prize–winning novel *The Road* are two recent examples of films that are charged with apocalyptic or post-apocalyptic sentiment as they tell their tales of total environmental destruction and its aftermath. While these two films are very different in terms of content and style, they are similar in their depiction of ecocide as end-time scenario and global human tragedy. Such a falling back on apocalyptical narrative, maintains Ursula Heise in her recent book *Sense of Place and Sense of Planet: The Environmental Imagination of the Global*, is a common feature in narratives that deal with global environmental risk scenarios (such as global warming) and their possible catastrophic consequences. New or alternative modes of imagining such scenarios on a global scale, Heise argues, are only just in the making.

One such alternative mode of narration, I will argue in the following, is presented in Andrew Stanton's animated blockbuster *WALL-E*, in which the long-term effects of total global ecocide make for a terrific comedy and biting satire. Not that things look all that rosy in the film's imagined twenty-eighth century: Set 800 years in the future, *WALL-E* shows a grim present in which, as a result of human over-accumulation and carelessness, the entire planet has been transformed into an enormous toxic trash heap, whirled up now and then by raging storms of debris. While the future effects of global warming do not play much of a role in the film, the desolate state of planet earth is clearly shown to be a direct consequence of mindless human action. The only two moving creatures left on the now brown planet are a nameless roach and an endearing, anthropomorphized robot called WALL-E, who passes his days compacting trash and his nights in a truck where he collects memorabilia of human life on earth. William Cronon has famously argued that for many Americans wilderness is imagined as "an island in the polluted sea of urban-industrial modernity, the one place we can turn for escape from our own too-muchness" (69). In *WALL-E*, this precious island of wilderness has long ago been submerged by human too-muchness. There is no wilderness—and in fact no nature—left on planet earth in the film's dystopic present. We only ever see what seem to be the remnants of New York City, but it is made clear that New York represents the planet as a whole, which is now bare of human, animal and plant life and completely engulfed in human trash.

In a way, the film is thus taking to the extreme Bill McKibben's thesis of "the end of nature," showing us a future in which not only the environment of specific wilderness areas or supposedly independent ecosystems have been destroyed as result of human interference but the entire natural environment of planet earth.[1] However, *WALL-E*'s earth is not only post-natural; it is also quite literally post-human, because not a single human being is left on the planet. What is left of humanity has sought refuge on the Axiom, a gigantic and thoroughly synthetic space ship that drifts aimlessly through outer space. What we find in *WALL-E* is thus an extreme de-centring of the human subject on a post-ecocidal earth that cannot any longer sustain animated life as we know it and that is thus now more suited for robots than for animate beings. That these robots are deeply anthropomorphized with regard to their emotional make-up but do not share any other biological human needs makes them the perfect heroes in the post-natural world of the movie. And even in the second part of the film, when WALL-E encounters the surviving humans on the Axiom, the human subject remains de-centred and the robots take centre stage, making it eventually possible for humanity to return "home" to rebuilt their desolate planet.

If much of this sounds depressing, it is not so in the film. In part, this is due to the fact that its title-giving main character is arguably the best non-human comedian Pixar Animation Studios has ever brought to cinema screens. And when WALL-E meets EVE—a sleek Extra-terrestrial Vegetation Evaluator robot sent back to earth by the crew of the Axiom—the comedy is spiced up by a romance which itself at times is hilariously funny. Using the tools of cognitive film theory, I will explore in the following how WALL-E's innocent, at times clumsy, but always ingenious and in the end successful attempts to win the heart of the somewhat belligerent beauty secure the emotional engagement of the audience and provide the much-needed comic relief in the face of global environmental disaster. Throughout much of the film, WALL-E is a romantic version of what Robin Murray and Joseph Heumann, following Joseph Meeken, have called the "comic eco-hero": a bumbling, kind-hearted and somewhat naïve main character who only accidently acts "heroic," doing so for mostly ulterior motives but with the full support of his community.[2] WALL-E becomes an eco-hero only by accident, and his saving of what is left of humanity from its eternal exile in space is in fact not much more than a mere side effect of his romantic intentions. He is a highly sympathetic character who wins the spectator's allegiance not only for his cuteness and kind-heartedness but also for his immense (in fact superhuman) courage in the face of even the most tremendous odds.

Combining highly aestheticized images of total environmental devastation and human degeneration with a funny and emotionally engaging love story, *WALL-E* turns eco-trauma into post-apocalyptic romantic comedy—a fact that makes its biting critique of American-style consumerism palatable for a mass audience.

ENVIRONMENTAL DISASTER, EMOTIONAL ENGAGEMENT AND COMIC RELIEF

Depictions of human waste and toxic landscapes have been part of the American literary and filmic landscape at least since Rachel Carson's seminal *Silent Spring*, and since the early 1980s a shifting relation to the natural environment is reflected in a large number of cultural texts. As Cynthia Deitering points out in her essay on the American post-natural novel, "during the 1980s we began to perceive ourselves as inhabitants of a culture defined by its waste" (197), and the resulting "ontological transformation," to use Deitering's term, was not only reflected in literature but also in American film. There is, in fact, an even longer tradition of eco-disaster films, originating already in the 1950s, which up to the 1980s "were a serious affair" (109), as Robin Murray and Joseph Heumann observe in their recent book *Ecology and Popular Film* (2009). Since the 1980s, however, Murray and Heumann maintain, "eco-disaster films have come of age," and disaster film genre conventions are now well enough known to be played for laughs in a satire or parody (*Ecology* 110). Interestingly, however, these more humorous approaches do not necessarily weaken the films' potential to convey a serious message. On the contrary, according to Murray and Heumann, "the same environmental message [can] be presented at least as effectively in a comedy in 1985 or today as it was in a more serious science fiction film in 1954" (*Ecology* 110). Like other comedies, comic eco-disaster films and comic eco-dystopias can very effectively comment on problems, dangers and grievances in their contemporary societies using humour, dramatic irony and hyperbole to get their critique across. This is particularly interesting, because like Deitering, Murray and Heuman notice a second, parallel shift that takes place as eco-disaster films mature: They increasingly move "away from the 'nature attacks' vision to one in which humans attack the natural world" (*Ecology* 111), bringing large-scale environmental destruction upon themselves and others. This, of course, is exactly the scenario depicted in *WALL-E*.

Following Maurice Yacowar, Murray and Heumann list three types of comic disaster movies: films that have classical "happy endings," films that satirize the disaster they depict and films that parody established genre conventions (see *Ecology* 111). *WALL-E* arguably combines all three elements in its plot line, as it unfolds its humorous vision of a post-ecocidal world before our eyes. Despite its lighthearted approach to ecological disaster, the film was lauded by a large number of reviewers for its potential to raise environmental awareness. Reviewing the movie for the *Los Angeles Times*, Kenneth Turan writes that *WALL-E* "can't help but send out a powerful and even frightening environmental message," and that the film's "dystopian vision . . . of what the perils of consumer excess have in store for the planet is unnerving without trying too hard" (Turan). And Michael Phillips, in his review for the *Chicago Tribune*, even argues that *WALL-E* presents a grimmer future for our planet than Al Gore's *An Inconvenient Truth*, and that "its strains of comedy and pathos" make the film nothing less than "a transforming experience" (Phillips). This probably overstates the case somewhat; however, Phillips certainly has a point. At their best, suggest Murray and Heumann, comic disaster films can provide "a space in which ecological problems . . . can be examined and scrutinized through . . . humor that intensifies an environmental message while minimizing didactic and pedantic proselytizing that a more serious approach might foster" (*Ecology* 113). As a post-apocalyptic romantic comedy, *WALL-E* is not in the least interested in didactic proselytizing, but instead relies on the power of breathtaking animated images coupled with a good dose of slapstick humour and a deeply romantic love story.

The importance of the love story, and the emotional engagement that it engenders in the viewer, cannot not be underestimated in its importance for the overall effect of the film. Movies, explains the cognitive film theorist Noël Carroll, "are objects that are well constructed to elicit a real emotional response from our already existing emotion systems" (23). It is indeed quite remarkable, Carroll notes, "to what extend . . . emotions . . . provide the cement that holds our attention on the popular movies we consume," often constituting "the most intense, vivid, and sought-after qualities available in the film experience" (23–24). According to Carroll, the emotions we experience when watching a tear-jerking melodrama, a scary horror film or an arresting suspense drama are not really different in kind from the "real" emotions we have in our everyday lives. What's different is that in life "our emotions have to select out the relevant details from a massive array of largely unstructured stimuli" (Carroll 28). When we watch a film, on the other hand, "the filmmakers have already done much of the work of emotionally organizing scenes and sequences for us through the ways in which [they] have foregrounded what features of the events in the film are salient" (Carroll 28).

Films, Carroll argues, *guide* our emotions while we watch them; they, as film theorist Greg Smith puts it, continually "offer [their audiences] invitations to feel" in certain ways (11). We can accept those invitations and "experience some of the range of feelings proffered by the text;" or we can reject them (Smith 11). If we choose to accept them, the film becomes a full body experience, because emotions, as Carroll explains, involve both cognition and "feelings," the latter being "sensations of bodily changes," like involuntary muscle contractions, the welling of tears, or the impulse to laugh

(24). Drawing on the insights of cognitive psychology, film theorists like Carroll argue that there is really no division between rational and emotional cognition, and that "the audience's faculties of cognition and judgment" are in fact of central importance in "the process of eliciting an emotional response to film" (41). In order to be successful, a film must shape what Carroll calls the *emotive focus* of the audience, determining the "way in which the emotional state of the viewer fixes *and* then shapes her attention" (31). This means that the emotional state of the viewer directly influences how she interprets and responds to individual scenes and sequences, and also how much she *cares* about the outcome of the story.

Central to the emotional engagement of the audience in narrative films is, generally, the main character or hero of the story. According to Carroll, different film genres rely on different combinations of emotional responses in their storytelling, but all of them are tied to the films' main characters. In romantic comedies, in which romance is presented in a comedic style, the emotional response is elicited by the protagonist's struggle to win the heart of his or her "true love." According to Bill Johnson, romantic comedies are so vastly successful with their audiences because, like romances, they embrace the maxims that "(1) true love does exist, (2) there's someone out there just for us, and if we could only find them, we would experience true love, and (3) romance can overcome all obstacles" (Johnson). The major difference between romance and romantic comedy, according to Johnson, is that the latter approaches its topic with a certain "light touch," making it all the more enjoyable to watch its hero overcome all (comic) obstacles (Johnson). This is why the protagonist of *WALL-E* is of such central importance for the emotional engagement of the film's audience.[3] Despite the fact that he is not human, WALL-E is deeply lonely at the beginning of the movie. Once he has met EVE, however, he enthusiastically and unflinchingly embraces the belief that he has found his true love, facing and overcoming every obstacle that stands between him and the fulfilment of his true love—and saving humanity from its post-natural dystopian fate in the process.

WALL-E AS POST-HUMAN LOVER AND COMIC ECO-HERO

WALL-E opens with a shot of the earth from space and to the happy tunes of "Put on Your Sunday Clothes," a song from the 1964 musical *Hello Dolly*. Already in these first seconds, however, we get a sense that something might be wrong with the planet, as our view is clouded by a thick layer of space-junk that encircles the earth. The camera then zooms down to the surface of the earth, flying over dark mountains of garbage, wind turbines and the remainders of nuclear power plants, and then offering a bird's eye view on towering skyscrapers built of trash in the golden afternoon light of a beautiful and seemingly hot day. The song fades out, leaving the audience with an ominous silence, and as the camera gets closer to the surface, we realize that there is no life on the streets of this city, and that they are instead covered with heaps of trash. Just as that realization begins to sink in, though, we hear the cheerful song again, at first barely audible. It now clearly emanates from the only moving creature in the setting, which turns out to be a small caterpillar-like robot who is busy compacting ridiculously small amounts of the omnipresent trash and stacking it up in cubes. On the robot's rusty and dented surface we read the letters WALL-E, which, as we will learn later, is an acronym for Waste Allocation Load Lifter Earth Class. As WALL-E notices the sun setting against the city's morbid skyline, he stops the music, picks up his tool box and his little friend, the roach (which is the only animal living on the planet), and rolls back through the waste-land of the city to his "home"—an old garbage truck.

What is perhaps most staggering in these first scenes of *WALL-E* is the sharp contrast between the cheerful-sounding music score, the beautifully lit images, the cute interactions between robot and roach and our more or less slow realization that all of this is actually set on a *dead* planet. The skyscraping towers of trash bathed in hazy sunlight have an almost sublime quality to them, and even as we encounter the absurd accumulation of one single corporate logo—that of the mega-corporation Buy-N-Large—all over the city, there is a certain aesthetic pleasure in the sheer colourfulness and artistic mastery of the images. This aesthetization of environmental catas-trophe seduces the viewer to ignore the grave environmental implications of the scene presented, and this negligence arguably is further assisted by the hero's indifferent behaviour. After all, as a robot WALL-E does not really need a clean city or even a biosphere, and neither does, apparently, his little friend the roach.

WALL-E's Sisyphus-like daily routine gives us a glimpse into how he has spent the past seven hundred years. Running on solar power and replacing his parts from defunct robots of his own kind, the passing of time is of little importance to the hard-working automaton, and he clearly does not care all that much about the dreadful state of the world he works in. This might be expected from a non-sentient being; however, from the very beginning of the film WALL-E is clearly anthropomorphized. Not only is he ticklish and able to laugh, he also interacts socially with his little friend the roach and seems quite enamoured with the human gadgets that he collects in his adopted home. "After 700 years of continuing his directive, i.e. cleaning the planet," the film-makers explain on their website, WALL-E has "developed one little glitch, a personality." This personality, however, does not include concern about the natural environment. Rather, it involves the development of inter-personal affect, about which WALL-E learns from a videocassette of the 1969 movie version of *Hello, Dolly!* which he keeps as one of his treasures. The musical as a whole and especially the lyrics of two of its songs—"Put on Your Sunday Clothes" and "It Only Takes a Moment"—play a crucial role in the movie, as WALL-E learns from watching the 850-year-old videotape about the importance of holding hands, and, by extension, human affection and companionship. Director Andrew Stanton has said in an interview that he "happened to have read somewhere that holding hands is the most intimate public display of affection, which led to the idea of WALL-E learning that action by watching the movie" (Stanton, "WALL-E Meets"). Imitating the gesture with his own two mechanical hands, however, the little robot is unable to understand what it means.

This changes when WALL-E meets EVE, the stylish and somewhat bird-like robot that is spit out by a futuristic space glider one day. While, as Vivian Sobchack has pointed out, "WALL-E is figured as primarily mechanical—familiar not only in his visible accomplishment of routinized workhorse labour but also in his material vulnerability to age, decay and obsolescence" (380), EVE's elegant and streamlined appearance mark her as member of a future generation of robots.[4] Just like WALL-E is unmistakably gendered male, EVE comes across as a somewhat belligerent female, and for a while WALL-E has his hands full avoiding being grilled by her ever-ready laser cannon. EVE also has her softer side, though, especially after WALL-E has introduced her to his little museum of consumer trash. While the two robots do not share a language and the only dialogues they exchange in the first thirty minutes of the film are the utterances of their respective names, WALL-E is clearly taken aback by EVE's streamlined elegance and soon—in love. The utter devastation of planet earth is, at this point, almost forgotten and the viewer has become so used to the aestheticized images of the brown landscape that any initial sense of shock is likely to have given way to the delightful enjoyment of WALL-E's somewhat inept courtship.[5] The film has become a boy-meets-girl story, a terrific romantic comedy which heavily relies on the strange beauty of its computerized cinematography and the superb acting qualities of its two animated main characters. As Carl Plantinga reminds us "the viewer's relationship with a favored character is one of sympathy, together with an assimilation of the character's situation" (103), and, like other romantic comedies, *WALL-E* invites us to sympathize with its two lovable main protagonists, and to feel solidarity for their interests and goals.

The turning point of the story occurs when WALL-E shows EVE the little green plant that he has found and that he keeps in an old boot in his truck. Like everything else she encounters, EVE scans the little green thing in front of her—the only green or greenish thing, in fact, that is to be seen in the film—and, having processed the data, she immediately grabs the plant, locks it in her interior and shuts down all of her communication systems. WALL-E is shocked at this sudden change, staring dumb foundedly at the little green sign in shape of a plant that is now blinking on EVE's white chest. In the following hours and days he tries everything he can think of to get EVE out of her sudden indifference and disregard. He even tries to force his little mechanical hand into the little crack in her egg-like shape where he knows her hand to be, hoping to produce the kind of emotional bond that he has seen in *Hello, Dolly!* But to no avail. EVE does not react to any of his advances, waiting apathetically for the arrival of her space shuttle. When that finally arrives and EVE leaves without even saying good-bye, WALL-E cannot bear the prospect of being alone and without his love again, and so he holds on firmly when the shuttle takes off to return to the Axiom. This romantic hero, we realize, goes to extremes in order to stay close to his beloved girl.

Life on the Axiom, where the rest of the film is set, could easily be read as another dystopic preview of a possible post-ecocide future, and it most definitely is a scathing critique of American-style consumer capitalism. Extremely obese humans spend their lives in floating armchairs, virtual video screens so close to their faces that they hardly notice their brightly coloured surroundings—let alone other humans. Time has become a meaningless concept on board the Axiom, too, because for the incapacitated passengers of the fully automated ship, life has become an eternal monotony of liquid foods, into-your-face commercials and pointless colour changes of their identical clothes. All deeds are done by robots which are controlled—like everything on the Axiom—by Buy-N-Large, the dictatorial mega-corporation which also played a defining role in the mindless over consumption that led to the destruction of the earth's biosphere. Retrospectively, we learn that Buy-N-Large gained more or less complete control over global mass consumerism throughout the twenty-first century, until the planet was finally covered by trash in 2105. In an attempt to resolve the situation, what was left of humanity—apparently almost exclusively Americans—was evacuated on fully automated space ships provided by Buy-N-Large to spend the next five years on a cruise while trash compactor robots would clean the planet. After those five years had passed, however, it became clear that the planet was too barren and too toxic to support life now or in the next few centuries, forcing humanity to remain in space until further notice.[6]

While the film's depiction of humanity's fate in the twenty-eighth century would be rather chilling if taken seriously, it is so satirical and grotesque that it invites a much more light-hearted reaction from the audience. Murray and Heuman, as I have pointed out earlier, argue that the strength of the comic eco-film lies in its potential to offer a space in which ecological problems can be examined through humour. In *WALL-E* it is not so much the ecological problems that are funny as the degraded humans who have mindlessly destroyed the planet and condemned themselves to a senseless life in outer space. In his review for *Entertainment Weekly*, Owen Gleiberman calls the film an "eye-boggling future-shock adventure" in which humans have "'evolved' into hilariously infantile technology-junkie couch potatoes" (Gleiberman), and there can be little doubt that in its portrait of future humanity, *WALL-E* dips deeply into the tool box of satire. Satire, as Geoff King explains, is comedy with a "political edge" (18), and while there is a whole other paper waiting to be written about the stigmatization and ridicule of obesity in the film, one can hardly overlook the fact that what is left of humanity in 2800 lives the American nightmare. The women and men we meet on the Axiom show some diversity in their skin colours (white and black mostly), but other than that they are homogenized—and that means Americanized—to the extreme. They all speak American English, wear the same clothes, ride on the same gliders and drink the same liquids. They all have lost their capacity to move (due to obesity and severe bone loss), to communicate or to do anything else but consume. Ironically, however, the founders and CEO's of the corporation that runs their lives have died long ago, and it is now the corporate computer AUTO who controls everything on the Axiom. AUTO is both an homage and parody of HAL in Stanley Kubrick's *2001: A Space Odyssey* (1968) and, as in Kubrick's classic, the computer turns out to be a danger to the ship's human passengers. Having received his final directive centuries ago when the clean-up project on earth was considered a failure, AUTO is programmed to keep up the status quo on the Axiom at all costs and thus humanity in space forever, irrespective of new or different circumstances.

With EVE's return and WALL-E's arrival on the Axiom, however, those circumstances change dramatically. Not only does EVE bring the evidence that there is plant life again on planet earth. She also brings—unwittingly—WALL-E, who in no time, quite literally, turns life on the ship upside-down, causing the humans to eventually see and talk to each other. This, however, is merely a side effect of WALL-E's actual quest. His only reason to be on the Axiom is his wish to be close to EVE. Had she not seen the plant and remained with him on the trash-covered earth, he would have been perfectly happy there and humanity forever drifting in space. This makes him a typical *comic* eco-hero in Joseph Meeker's understanding of the term. Unlike the tragic eco-hero, Meeker explains, the comic eco-hero does not necessarily act out of high-minded idealism and may even be "weak, stupid or undignified" (158). Because he does not have heroic intentions, Murray and Heumann explain with recourse to Meeker's definition, the comic eco-hero "tend[s] to bumble and require[s] a community of allies to succeed" (*Ecology* 111). WALL-E is exactly such a bumbling hero, kind-hearted and curious, but also naïve and a little timid. Compared to EVE's fearless and potentially dangerous beauty (something of a robot version of Angelina Jolie), WALL-E's boyish charm seems wonderfully human but not exactly heroic. And as with most comic eco-heroes, his heroic deeds are more fortuitous than actually intended. While he certainly cannot be called stupid, he often lacks understanding of a given situation and he really has no clue about the tragic fate of planet earth or that of the humans on the Axiom.

The only person who, at some point, begins to realize and formulate the ethic dimension of the situation of both the planet and humanity is the captain of the ship. While he shares all physical features with the rest of the humans on board the Axiom and has spent most of his life in blissful and careless drift, he awakens from his prolonged stupor when seeing the images, explanations and definitions that the Axiom's computer system offers him after he has given it some crumbs of the earth's soil to analyze that he found in WALL-E's hand. When he later reviews the visual records that EVE has brought back from the present-day earth he is deeply shocked, and because he—unlike WALL-E—has the power of human speech, he is also the one who communicates the implications of the disaster to the audience. "You just need someone to look after you, that's all," he says to the little green plant that EVE brought him after he has given it some water—and realizes immediately that that is not only true for the plant but for the planet as a whole. Understanding humanity's responsibility for the horrific ecological destruction of the earth and fascinated by the idea of planting and harvesting crops, he vows to return and restore the planet. But even though the good-natured captain in a way is the true "eco-hero" of the story and although he plays an important role in the liberation of the humans on the Axiom, he never even gets close to becoming a central protagonist—that is a privilege exclusively reserved for the robots.

WALL-E, however, as the main and title-giving character, has little or no awareness of the sad ecological fate of the earth or the destiny of the humans.[7] The true dimensions of this vast eco-trauma are incomprehensible to him and they thus have no bearing on his actions. While he is the one who first found the little green plant back on the earth, and was fascinated enough by it to keep it, he does not seem to see any difference between it and any other gadget he collects (in fact, it is a miracle that it survived in the darkness of his garbage truck). Even as he encounters human life in space he does not understand any relations between them and the place he came from—how could he? And thus his feelings for EVE, and his determination to follow and protect her, are the sole reason why he becomes the (accidental) eco-hero of the film. His tremendous dedication to her allow him to go beyond himself in his desperate fight against AUTO, and with the support of a horde of malfunctioning robots—and the captain—he manages to win this fight, and EVE in turn falls in love with him. None of this, however, is done for the sake of humanity or the betterment of the sad condition of the earth per se. WALL-E's part in the liberation of both robots and humans and the resulting return of humanity to restore the earth is at best something of a side effect to his personal quest for true love.

The film, then, does not ask its viewer to get emotionally involved with the fate of planet earth—at least not directly. Noël Carroll, we might remember, has argued that, in order to be successful, a film must shape the *emotive focus* of the audience, determining the "way in which the emotional state of the viewer fixes *and* then shapes her attention" (31). In *WALL-E*, the emotive focus of the audience is shaped in such a way that we want the little hero of this romantic comedy to succeed, that we hope he will be able to win his beautiful EVE and live with her happily ever after.[8] As Plantinga explains, character sympathy "is pleasurable in itself, but it also ensures strong emotional responses, because when the audience cares deeply about a character, it also has deeper concerns about the unfolding narrative" (111). One of the emotional promises of romantic comedies is that the couple will be united in the end, and *WALL-E* is no exception.[9] In what can easily be criticized as a sentimental and all-too-American ending, the captain eventually steers the Axiom back to earth, to clean up the planet and start a new era of ecologically sustainable human life. *WALL-E* is, after all, family entertainment, and not by coincidence produced and distributed by Walt Disney Pictures.

The fact, however, that the emotional ending of the film includes humanity's awakening out of its century-long apathy and its sudden realization of the prime importance of both interpersonal relationships *and* a respectful relation to the earth's natural environment, cannot be ignored. While the main plot of the film concentrates on the love story between WALL-E and EVE, the happy ending includes not only their personal happiness but also that of humanity and the earth, who, we are led to believe, will have a healthier relationship in the future. As Murray and Heuman point out in their *Jump Cut* article, "*WALL-E* draws on nostalgia to strengthen its argument that not only has humanity destroyed earth, but that humans—with the help of the robot left to clean up the mess—can and should restore it to its more natural previous state" ("*WALL-E*").[10] And the fact that the film combines the happy ending of its love story with this human return to earth increases our emotional involvement in the latter.

GREEN HOLLYWOOD OR UNINTENDED SIDE EFFECTS?

This analysis sheds some doubt on director Andrew Stanton's repeated claims that he did not actually *mean* to create an environmental movie and that the ecological message that so many reviewers and viewers see in the film is an unintended side effect at best. In an interview with *New York Entertainment*, Stanton insists that his sole interest in making the film was to create a touching love story around the last robot on earth, and that he did not have "a political bent or ecological message to push" (quoted in Simon). Admittedly, the film's merchandise-spewing marketing campaign suggests that Disney-Pixar was not planning to make an environmentally friendly film, either.[11] *WALL-E* thus hardly counts as what David Ingram has called "film vert"—films with an openly environmentalist message or advocacy—nor can it boast to be a carbon-neutral production, like *The Day after Tomorrow* (2004) or *Syriana* (2005). If it is nevertheless a successful environmental movie, this seems to be more due to happenstance than any intention on the side of the film-makers. While eco-trauma and its disastrous effects clearly are pervading themes throughout the film, Stanton has made clear—at the Academy Awards Ceremony, where he received the Award for the Best Animated Feature—that the main idea that he had in making the film was a different one:

> I think a lot of people attach a little too specifically to the ecological aspect or the complacency aspect of humanity. But I use those as

devices to focus on the biggest issue, which is people caring about one another. People connecting with one another. Whether that's literally love between two characters like robots or just you acknowledging that your neighbors right next to you as opposed to being blocked between a cell phone or something.

(Stanton, "Press Transcripts")

Like his hero, then, Stanton claims to be an involuntary eco-hero at best, a fact that has puzzled quite a few of his interviewers. However, Stanton is at least willing to acknowledge that the environmental aspect of humanity's complacency as it is depicted in his film is "going to be the cause, indirectly, of anything that happens in life that's bad for humanity or the planet" (quoted in Simon). Whether it was intended or not by the film-makers, however, if we judge the film by the responses of critics, journalists and audiences, *WALL-E* seems quite powerful in its depiction of eco-trauma and quite successful in raising awareness about the possible ecological consequences of human action.

And perhaps even more than that. Murray and Heumann, at least, suggest that "laughing about the environment and its degradation may not only stimulate awareness; [it] might also point out a path toward change" (*Ecology* 113), and this might be even more true for *WALL-E* than for other eco-comedies. After all, one should not forget that the target audience of the film is not necessarily the seasoned environmental activist or other longstanding members of the ecological choir. As film critic Jason Coffman points out, "for adults, the film's message may seem a little obvious and overly preachy, but for a My First Anti-Corporate Message film, it will doubtless find a lot of young fans who take its message to heart" (Coffman). Stanton, on his part, has said that the film is a "perfect metaphor for real life" as at least as the younger generations in the industrialized North live it today (quoted in Fritz). And the fact that *WALL-E* is such a pleasant and entertaining film to watch, and as an animated picture about robots so seemingly far away from our present-day concerns, might even help in conveying its more subversive messages. After all, film theorist Jean-Loup Bourget has insisted that even "escapism can . . . be used as a device for criticizing reality and the present state of society" (52). Depicting a future world in which humans have ceased to be of any importance and relating the consequences of eco-trauma directly to past human over consumption, even a film as escapist as *WALL-E* might have some real world effects. Having seen the film, and being moved by its touching story and brilliant images, the more sobering news that it would take three to five planet earths to sustain current European and American lifestyles is likely to register much more forcefully with young viewers.

WALL-E thus does offer one possible alternative mode of imagining scenarios of environmental risk and ecological trauma on a global scale. Seducing its audience with an emotionally engaging combination of slapstick humour, satire and romantic comedy, the film confronts it at the same time with the desolate landscapes of a post-ecocidal planet that is the direct result of anthropogenic environmental impact and looks at least as barren and lifeless as the one presented in John Hillcoat's post-apocalyptical tale *The Road*. And while the latter film, too, focuses on a "love story" between two humans—in this case father and son—the two movies could not be more different in their approach to environmental trauma. When at the end of *WALL-E* the two robots are finally lovingly holding hands to the sentimental tunes of *Hello, Dolly!* and images of humans and robots working together to rebuild their planet accompany the end credits, at least the younger ones among the audience may decide that they need to do something to prevent the destruction of the planet they currently live on. While time might not matter much after total global ecocide has taken place, they might have learned from *WALL-E* that it matters a lot if we want to prevent it.

NOTES

1. In *The End of Nature* (1989), McKibben argues that through its mismanagement of the earth's resources, humanity is inviting the end of the earth's biosphere and therefore its own existence as well as that of millions of other species. "By the end of nature," McKibben makes clear, "I do not mean the end of the world. The rain will still fall and the sunshine, though differently than before. When I say 'nature,' I mean a certain set of human ideas about the world and our place in it. But the death of those ideas begins with concrete changes in the reality around us—changes that scientists can measure and enumerate. More and more frequently, these changes will clash with our perceptions, until, finally, our sense of nature as eternal and separate is washed away, and we will see all too clearly what we have done" (7).
2. In their 2009 article on *WALL-E* in *Jump Cut*, Murray and Heumann even argue that WALL-E starts out as a tragic hero who, through internalizing "the messages of the artifacts he collects" slowly gains characteristics of the comic hero, evolving "from a machine to a more humanoid (and comic) android" (Murray and Heumann 2009).
3. Offering a breakdown of *Sleepless in Seattle*, Johnson explains that the film is based on the simple premise that "true soul mates can find each other in spite of any obstacles to their being together. Note how this premise applies not just to specific characters, but to an idea about romance that the story acts out dramatically" (Johnson). Pretty much exactly the same premise is at the heart of *WALL-E*, and just as in *Sleepless in Seattle*, the two main characters eventually overcome all odds to be happily together ever after. The fact that in this case the two "lovers" are not humans but anthropomorphized robots makes their struggle for true love even more amusing—and also more touching.
4. Vivian Sobchack argues that these computer-generated images remind us forcefully of the "death" of classical photochemical cinema while at the same time paying homage to it: "Historically and formally, WALL-E's simulations pay cinephilic homage to photochemical cinema—this not only through its haunting computergraphic 'cinematography' (Roger Deakins was a consultant) but also through its remarkable, dialogue-free, 'silent' first act, its explicit modeling of its quasi-mechanical protagonist on Buster Keaton and Charlie Chaplin, and its often hilarious citations of science fiction movies both classic and contemporary" (379).

5. In fact, a number of critics have pointed out that the first thirty minutes of the film, in which the story is told almost entirely through visuals and implied discourse, are by far the best.

6. In its depiction of a lifeless planet, *WALL-E* sometimes reminds of Douglas Trumbull's 1972 science fiction film *Silent Running*, only that in the case of Trumbull's film it is the earth's forests and not the humans that wait in space for their eventual return. If anything, the future of humanity as it is depicted in *WALL-E* is more frightening since here the earth cannot sustain human life in any way and since it takes seven hundred years for plant life to return after total global ecocide. An interesting development in *Silent Running* in this context is that at the end of the film the space-born greenhouse domes, containing the last remaining plant life of planet earth, are drifting alone away from earth and deep into space, left in the care of non-human drones.

7. Astonishingly enough, there are still babies, which are nursed and educated by robots, but given the physical and mental state of the Axiom's passengers; we can only assume that these children—in *Brave New World*–style—are produced in laboratories with the aid of sophisticated reproduction medicine.

8. In a 2009 article on *WALL-E* in *Jump Cut*, Murray and Heumann argue that *WALL-E* actually evolves from a tragic character to a comic character in the course of the film.

9. Murray and Heumann discuss the protagonist of *Eight-Legged Freaks* (2002) as a typical example for a comic eco-hero.

10. Murray and Heumann point to the film's three-part structure, which, in their view, moves from the establishment of earth as an inhospitable setting for human and nonhuman life, to the leaving of earth on an evolutionary journey, to the return to earth with the intention to transform it into a home. "Instead of highlighting a tragic narrative," the two scholars argue, "ultimately *WALL-E* shifts to a narrative that is embedded in the comic and communal, rather than tragic and individualized notions of species preservation found in the tragic evolutionary narrative of *The Odyssey* and of 'early Darwinism'" (n.p.). Those "tragic evolutionary narratives support extermination and warfare rather than accommodation, the results on display in *WALL-E's* opening shots" (Murray and Heumann n.p.).

11. In an online piece on CHUD.com (Cinematic Happenings Under Development) Kevin Faraci accuses the film-makers of hypocrisy and complains that at a press conference promoting the film "the room was stuffed with what seemed like a hundred or more tie-in products ranging from *Wall-E* branded plastic Crocs (with tire tread patterns on the soles) to plastic *Wall-E* action figures to *Wall-E* branded clothing and bed sets and drapes" (n.p.). Other commentators, however, have stated that the merchandising campaign of *WALL-E* was not particularly extensive or successful, compared to other Pixar productions, like, for example, *Cars* (2006).

WORKS CITED

2001: A Space Odyssey. Dir. Stanley Kubrick. Metro-Goldwyn-Mayer, 1968. Film.
2012. Dir. Roland Emmerich. Columbia Pictures, 2009. Film.
Bourget, Jean-Loup. "Social Implications in the Hollywood Genres." *Film Genre Reader III.* Ed. Barry Keith Grant. University of Texas Press, 1986. 51–59. Print.
Carroll, Noël. "Film, Emotion, and Genre." *Passionate Views: Film Cognition, and Emotion.*" Ed. Carl Plantinga and Greg M. Smith. Baltimore: The Johns Hopkins University Press, 1999. 21–47. Print.

Carson, Rachel. *Silent Spring*. Boston and New York: Houghton Mifflin, 1962. Print.

Coffman, Jason. "Review WALL-E." *Film Monthly*. July 3, 2009. www.filmmonthly. com/video_and_dvd/wall-e.html. Web. November 2009.

Cronon, William. Ed. *Uncommon Ground: Rethinking the Human Place in Nature*. New York: Norton, 1996. Print.

Cars. Dir. John Lasseter. Pixar Animation Studios, 2006. Film.

The Day after Tomorrow. Dir. Roland Emmerich. Twentieth-Century Fox, 2004. Film.

Deitering, Cynthia. "The Postnatural Novel." *The Ecocriticism Reader: Landmarks in Literary Ecology*. Ed. Cheryll Glotfelty and Harold Fromm. Athens and London: The University of Georgia Press, 1996. 196–203. Print.

Eight Legged Freaks. Dir. Ellory Elkayem. Village Roadshow Pictures, 2002. Film.

Faraci, Devin. "The Devin's Advocate: Is Wall-E Environmental or Hypocritical?" *Chud.com*. June 23, 2008. www.chud.com/15280/the-devins-advocate-is-wall-e-environmental-or-hypocritical/. Web. November 2009.

Fritz, Steve. "How Andrew Stanton & Pixar Created WALL*E—Part II." *Newsarama*. July 4, 2008. www.newsarama.com/film/080704-wall-e-stanton-2.html. Web. November 2009.

Gleiberman, Owen. Review of *Wall-E*. *Entertainment Weekly*. June 26, 2008. www. ew.com/ew/article/0,,20209111,00.html. Web. November 2009.

Heise, Ursula K. *Sense of Place and Sense of Planet: The Environmental Imagination of the Global*. Oxford and New York: Oxford University Press, 2008. Print.

Hello, Dolly! Dir. Gene Kelley. Perf. Barbra Streisand, Walter Matthau, Michael Crawford. Twentieth Century Fox, 1969. Film.

Ingram, David. *Green Screen: Environmentalism and Hollywood Cinema*. University of Exeter Press, 2004. Print.

Johnson, Bill. "The Art of the Romantic Comedy." A Story Is a Promise. www. storyispromise.com/wromance.htm. Web. November 2009.

King, Geoff. "Spectacular Narratives: *Twister, Independence Day* and Frontier Mythology in Contemporary Hollywood." *Journal of American Culture* (1999): 25–39. Print.

McKibben, Bill. *The End of Nature*. New York: Anchor Books, 1990. Print.

Meeker, Joseph. "The Comic Mode." *The Ecocriticism Reader: Landmarks in Literary Ecology*. Eds. Cheryll Glotfelty and Harold Fromm. Athens and London: The University of Georgia Press, 1996. 155–169. Print.

Murray, Robin L. and Joseph K. Heumann. *Ecology and Popular Film: Cinema on the Edge*. Albany: State University of New York Press, 2009. Print.

_____. "WALL-E: From Environmental Adaptation to Sentimental Nostalgia." *Jump Cut: A Review of Contemporary Media*, Spring (2009). May 31, 2010. http://ejumpcut.org/archive/jc51.2009/WallE/index.html. Web. June 2014.

Phillips, Michael. "'WALL-E': 4 Stars! Pixar's Trash-Compacting Robot Will Be Collecting Hearts." *Chicago Tribune*. June 27, 2008. www.cincinnatibell.net/ movies_channel/reviews?movie_id=63804. Web. November 2009.

Plantinga, Carl. *Moving Viewers: American Film and the Spectator Experience*. Berkeley: University of California Press, 2009. Print.

The Road. Dir. John Hillcoat. Dimension Films and The Weinstein Company, 2009. Film.

Silent Running. Dir. Douglas Trumbull. Universal Pictures, 1972. Film.

Simon, Brent. "Pixar on 'Wall-E': Environmental Themes? What Environmental Themes?" *New York Entertainment*. June 26, 2008. www.vulture.com/2008/06/ pixar_on_walle_environmental_t.html. Web. November 2009.

Smith, Greg. M. *Film Structure and the Emotion System*. Cambridge and London: Cambridge University Press, 2003. Print.

Sobchack, Vivian. "Animation and Automation, or, the Incredible Effortfulness of Being." *Screen* 50.4 Winter (2009): 375–391. Print.

Stanton, Andrew. "WALL-E Meets Dolly." Interview. *EW.com. Entertainment Weekly*. July 14, 2008. www.ew.com/ew/article/0,,20211943,00.html. Web. June 2014.

———. "Press Transcripts. 81st Academy Awards." Interview. *Academy of Motion Picture Arts and Sciences*. November 10, 2009. www.oscars.org/press/transcripts/ Web. June 2014.

Syriana. Dir. Stephen Gaghan. Warner Bros., 2005. Film.

Turan, Kenneth. "Review *WALL-E*." *Los Angeles Times*. June 27, 2008. www.latimes.com/world/middleeast/la-et-walle27-2008jun27-story.html. Web. June 2014.

WALL-E. Dir. Andrew Stanton. Pixar Animation Studios, 2008. Film.

WALL-E Official Website. May 21, 2009. http://movies.disney.com/wall-e. Web. July 2014.

9 Eavesdropping in *The Cove*
Interspecies Ethics, Public and Private Space and Trauma under Water

Janet Walker

Right now I'm focused in on that one little body of water where that slaughter takes place. If we can't stop that—if we can't fix that—forget about the bigger issues. There's no hope.

—Ric O'Barry, *The Cove*

[T]he impact of a traumatic event lies precisely in. . . . [I]ts refusal to be simply located, in its appearance outside the boundaries of any single place or time.

—Caruth 1995, 9

The eyes of dolphins are located on the sides of their heads and may be "pooched out" to enable a full 180-degree field of view on each side or limited binocular vision towards the area below the snout or "rostrum." Dolphins have spherical lenses—as distinguished from the flattened human lens—that enable the animal to focus both in air and in water. After a few observations of a dolphin "speed-swimming upside down just below the surface" of the ocean, marine mammal researcher Kathleen Dudzinski realized that the animal was using binocular vision to see above the surface in order to catch flying fish the moment they reentered the water (Dudzinski and Frohoff 23). Even more developed is the dolphin's acoustic sense of echolocation, or the use of pulsed sounds or clicks to locate and investigate objects or features of the environment.

Dudzinski and her collaborator Toni Frohoff make a point of conducting their research under water, where dolphins as small cetaceans spend about 99 percent of their time, instead of from land or boats, the surface vantage point from which most of the existing scientific data about dolphin behaviour has been gathered. They call their practice "eavesdropping," and their "Etiquette for Interacting with Dolphins" includes not touching, chasing or feeding the creatures, but rather watching, listening, diving down and ceding them their space (166–68). Following the lead of these dolphin researchers, I seek to watch and listen to dolphins under water—while ceding them their space—through an improvised environmental media approach to *The Cove*

(Psihoyos 2009), a recent activist documentary about the yearly slaughter of dolphins in a "killing lagoon" formed by the coastline around Taiji, Japan.

Working with long-time pro-dolphin activist Richard "Ric" O'Barry, the Oceanic Preservation Society (OPS) has created the film to help put a stop to this annual hunt, in which a limited number of animals are culled for dolphinaria but the majority are driven around the point and killed. The subject matter is harrowing—a graphic title cites the estimated figure of 23,000 dolphin and porpoise deaths per year in Japan[1]—and the film a labour of compassion.

In academic settings, conversations have concentrated on whether the depiction of the coastal fishermen in this US film is ethnocentric, anti-Japanese or even racist, and whether its laser focus on dolphins and dolphin hunting in Taiji neglects what would be a more appropriate critique of the *global* seaquarium industry (including enormous profits being reaped in the West) and wild fish depletion. In other words, scholars have questioned the documentary ethics of the film's incursion into Japanese waters.[2] My own sense is that the film does acknowledge, but to a limited extent, the global network of dolphinaria, the machinations of the International Whaling Commission (which historically has not been recognized as having jurisdiction over small cetaceans) and catastrophic overfishing. These topics will be taken up later, along with a discussion of the film's nuanced distinctions between the handful of Taiji whalers and the broader Japanese public.

In any case, I want to begin by placing these critical intersectionalities in abeyance in order to concentrate on dolphins. What can learn by immersing ourselves in the film's dolphinophilia, even if we must breach it, in the final analysis, with our own transgressive criticality?[3] And since *The Cove*'s ecology is one of captivity and slaughter, how might the film's focus on dolphins enable us to engage with trauma studies concepts in an interspecies context? Bridging between "eavesdropping" as a marine mammal research practice and "observation" as a documentary function or mode (Nichols 1991; Sobchack and Sobchack 1980), this chapter will explore how director Louie Psihoyos and company's campaign for filming in the cove—which operation is reflexively featured in the film itself—makes visible the capacity of documentary film not only to sense and to represent but actually to remap and remake the natural environment. This chapter will also investigate how the film's technological and aesthetic practices for sounding out animal voices make audible questions about the inclusion of dolphins subjected to violence as a group appropriately thought of in terms of traumatization, genocide, and/or Giorgio Agamben's articulations of the concept of "bare life." Informed by philosophies of critical human geography and ecocentrism as well as documentary and trauma studies, this analysis offers an interdisciplinary, or perhaps it would not be too bold to say ecosystemic, approach to the development of spatial and environmental media studies in the era of sea life depletion.

THINKING ACROSS SPECIES

The Cove builds its pro-cetacean argument on the basis of protagonist Ric O'Barry's critique of anthropomorphism, the attribution of human char-acteristics to non-human animals. Having made a career in the 1960s as a dolphin trainer with the Miami Seaquarium and then for the *Flipper* TV series about a Bottlenose dolphin[4] who befriends and aids the warden of a marine preserve (NBC TV, 1964–1967), O'Barry became an activist when Kathy, one of the dolphins "playing" Flipper, died in his arms. She "looked me right in the eye, and [audible whoosh as O'Barry sucks air into his mouth] took a breath and didn't take another one. And I just let her go and she sank straight down . . . On her belly. To the bottom of the tank." The next day, O'Barry was "in the Bimini jail for trying to free a dolphin." He goes on strongly to criticize the marine "captivity industry" with its *Flipper*-inspired dolphin shows for making Taiji hunting vastly more profitable than it would be if its object were solely the harvest and sale of dolphin meat, which "meat" has to be sneaked onto the market because of its high mercury content and because the whale meat sold for food in Japanese groceries has *not* traditionally included that of the small cetaceans.

A major problem, O'Barry tells us, is the popular tendency to misin-terpret the "dolphin smile" as a reflection of the animal's emotional state. This physiognomic feature that scientists know to be a "a by-product of the structure of the lower jaw" and "result of a morphological adaptation for sound reception" (Dudzinski and Frohoff 90) "creates the illusion they're always happy" (O'Barry), thus contributing to people's felt affinity with dolphins, and, in turn, to the popularity of dolphin shows and swim with dolphins programmes.

O'Barry and the OPS refute the avowed educational value of captivity-based programmes that supposedly teach children and adults to love dol-phins and other sea creatures. The truth of the matter is the other way around, expedition and technical director Simon Hutchins explained to students in my course on "Films of the Natural and Human Environment" and other audience members present at the panel discussion of the film (see Hutchins 2009). As with zoos, the dolphin shows are popular because peo-ple *already* love dolphins (the film includes a shot of a little girl hugging a stuffed dolphin next to a display shelf of the fuzzy marine mammals), and humans may benefit emotionally and socially from this felt connection to entertaining dolphins. But the benefits are far from mutual. An apt com-parison may be the English bulldog's debilitating respiratory problems that exist as result of breeding for a pushed-in snout more desirable to humans (Serpell 2005 cites Thompson 1996). In the case of dolphins, the cuteness factor exploited not only fails to raise consciousness for cetacean protection, but also serves as a contributing factor to their captivity. O'Barry mentions that the sound of tank filtration systems has been identified as a cause of stress and even death in dolphins as acoustic creatures.

The Cove is therefore critical of the *Flipper* TV show's anthropomorphizing tendencies,[5] exploited by O'Barry himself in his younger days. "I feel somewhat responsible . . . because it was the *Flipper* TV series that created this multi-billion dollar industry," he confesses (however grandly). We see O'Barry pulling a dolphin into a boat in a "behind the scenes" documentary from 1962 included on the DVD of the film: "She seems to sense that she has come home; that no harm will come to her now. She is safe," intones the narrator. Sandwiched between O'Barry's statement and the Orwellian publicity documentary, the show itself with its catchy theme song, animated intro and signature shot of an open-mouthed "laughing" Flipper reads as enormously exploitive of both the dolphin cast and its child audiences. The premise that Flipper lives in the wild and helps human "friends" of his own volition is efficiently revealed by *The Cove* as a pretense riding on the backs of the five female Flipper dolphins held in captivity in the Florida saltwater lake.

And yet, the film is actually structured around a contradiction: coexisting with its critique of anthropomorphism is a tendency to engage audiences in the lives of dolphins "by appeal to mental states similar to the ones we take to explain our own behavior"; that is, by appeal to the rhetoric of anthropomorphism (Mitchell 101). As O'Barry explains in the film, and presumably at his many speaking engagements around the world, at the time of its occurrence he understood—and still today construes—the *Flipper* dolphin Kathy's death as a suicide:

> She was really depressed. I could feel it. I could see it. And she committed suicide in my arms. I know that's a very strong word, suicide, but you have to understand, dolphins and other whales are not automatic air breathers like we are. Every breath they take is a conscious effort. And so they can end their life whenever life becomes too unbearable by not taking the next breath. And it's in that context that I use the word suicide. . . .

O'Barry and the film despair of dolphins who are "all stressed out," "freaked out," "really depressed" or suicidal. And of course, all of the dolphins who played Flipper had names they were given and that O'Barry continues to use. To the extent that the value of dolphin life is ascribed to their similarity to humans, the film may be seen as being caught up in the very anthropomorphic thinking it purports to reject.

But this is not to fail *The Cove* on its own litmus test requiring the rejection of anthropomorphism. In fact, I am intrigued by the possibilities of *The Cove*'s anthropomorphic proclivities for facilitating this conceptual move from a human-based to a more inclusive trauma studies perspective. The potential of anthropomorphic thinking for interspecies understanding has been addressed by scholars from various fields. As historian and philosopher of science Sandra Mitchell argues,

> anthropomorphism is neither prima facie bad or necessarily non-scientific. It can be both, but it need not be either. There has been a recent resurgence of interest in anthropomorphism, attributable to two developments—the rise of cognitive ethology and the requirements of various forms of expanded, environmental ethics.
>
> (100)

With careful attention to rhetorical logic, scientific experimentation and empirical findings, Mitchell explores claims of human-to-non-human causal isomorphism—that is, similarities in "neurophysiological structure, sensory apparatus, and so on" (110)—and finds them valid in certain cases. For example, "we are comfortable using the results of drug tests on mice to infer the consequences of those drugs on human biochemistry" (111). Toni Frohoff, for her part, embraces the term "zoomorphism" to emphasize what humans and other animals share as distinguished from how "they" are like "us," thus affirming the interspecies bond (4 November 2009).

From this perspective, we may sense *The Cove*'s array of felt affinities, observational findings and interspecies listening as sophisticated and useful for the identification of genuine commonalities. Kathy the dolphin watching the *Flipper* show (as we see, on a portable TV at the end of the dock) was able to distinguish between herself and the other Flipper players, O'Barry claims. Here he invokes the well-known mirror self-recognition argument that a dolphin knows when looking in the mirror that it is herself or himself that she or he sees. "They are self-aware the way humans are self-aware," he explains. The point is supported by scientific research and reporting: "[W]hen presented with a mirror, dolphins take the opportunity to check their teeth and body parts they can't normally see, like their anal slit" (Angier 2010). One of the experiments done to confirm self-recognition is marking dolphins on their sides with a (presumably non-toxic) pen, and seeing if they swim immediately to a mirror in the tank to turn and check what has just occurred on their bodies. They do. *The Cove* is very much in keeping with these scientific studies of dolphins and other cetaceans (Angier 2010; Leake 2010) in its emphasis on these creatures' individual self-awareness, feelings and emotions; non-human intelligence (but intelligence nonetheless); inter- and extra-group sociality; and ability to think about the future.

TRAUMA UNDER WATER

However, the film's recognition of dolphin sentience raises ethical quandaries in light of species differentiation and in the space of the cove. O'Barry states at the beginning of the film that "nobody has actually seen what takes place [in the killing lagoon] and so the way to stop it is to expose it." The dolphin hunters, including one man the film-makers dub "Private Space," would seem to concur. These members of the Taiji Fishermen's Cooperative (the name featured in their hand-lettered signs) seek to block the film-makers' access to the cove and the visibility of the hunt. The project team must therefore make a concerted effort to film the hunt, which effort involves eluding the Cooperative members either by physically getting around them or by filming clandestinely when they are not present.

The challenge was to present the full sensory environment of the slaughter as, in Psihoyos's words, a "three dimensional experience of what's going on in that lagoon." Hence the high-tech operation depicted reflexively and organized to culminate in two "missions" (first, "planting hydrophones" "to get the auditory experience" along with underwater footage, and second, keeping the sound theme alive, the "full orchestra").[6] Via the film itself, and in a more detailed fashion through the DVD's special features, we become privy to the necessary technical innovations: the use of gyro-stabilized high-definition cameras, protocols for remote operation and, of course, the fake rock housings in which cameras could be hidden above or under water created by artists at the world-renowned visual and special effects studio Industrial Light and Magic.[7] The film features numerous shots of equipment cases (at airport baggage check-in, pushed along on hotel caddies, lined up in a hallway outside hotel rooms), and multiple sequences showing the team innovating, testing, manipulating, hooking up and joking about the need to peel the OPS labels off equipment.

Strong capability for aural eavesdropping under water is referenced repeatedly in the film. On camera Hutchins mentions the use of hydrophones. In edited voice-overs, we hear "high tech sound devices put in underwater housing" being referred to: "I wanted to hear the dolphins in the lagoon, you know, how deep it was," explains Psihoyos. The sequencing in this passage of the film explains that world-class free diver Mandy-Rae Cruickshank was brought on board the project because she could descend in depth without the need for unwieldy, noisy ("clanky") scuba equipment that would compromise "stealth and speed" (Hutchins 2010).

In fact, Dudzinski's and Frohoff's notion of "eavesdropping" is defined specifically in relation to the aural dimension:

> When we . . . think of eavesdropping, we usually think of someone listening to a conversation without being observed—from behind a closed door, around a corner, or on a telephone extension. Remaining undetected allows us to learn much information from, and about, others.
>
> (64)

"But snooping on dolphins is a whole different story than eavesdropping on a sister's phone conversation," they continue. Dudzinski works with a "mobile video/acoustic system (MVA)" that she developed beginning in 1992 and that now takes the form of two hydrophones "located at the ends of a bar attached to the housing, where they are plugged into a stereo video camera" and "a third hydrophone, a digital audio recorder, and a circuit board to capture and record echolocation from wild dolphins" (81). Unlike Jacques-Yves Cousteau in *Le Monde du silence/The Silent World* (Cousteau and Malle 1956) mocking the vocalizations of a sperm whale ("I can hardly believe my ears, the giant squeaks like a mouse!"), Dudzinski and Frohoff are most emphatic that the whistles, clicks, screams and barks that dolphins make are a sophisticated form of communication.

From the perspective of the documentary studies, referencing the defining quality of the observational camera to disappear into the woodwork, we might note that in this case the "fly-on-the-wall" has sonar. Hearing the sounds of cetaceans may indeed remind us that they are sentient, acoustic creatures able to communicate with one another, if not with humans in a language we understand.[8]

One of the film's most powerful audio-visualizations, therefore, is a gruesome underwater seascape captured by a planted, submerged camera. In it we see waving plant life and a school of fish progressively enveloped by a red cloud of dolphin blood mixed with seawater. On the accompanying soundtrack, sounds that have been dredged up are made accessible to auditors on land. Over the roar of the sea through the hydrophone we hear the sound of dolphin whistles.

Back at the hotel, the team listen with rapt attention to their recordings of dolphins communicating with one another. "When [the dolphins are] in that killing cove and their babies are being slaughtered in front of them, they're aware of that. They can anticipate what's gonna happen to them," O'Barry states. Intercut with shots of the laptop and digital audio recorder we see medium close-ups of individual team members. They are listening not only to everyday dolphin vocalizations but also, silently, respectfully, to the sounds made by particular groups of dolphins in the Taiji lagoon while under attack. Their faces reflect the import of bearing witness, retrospectively. "It's an eerie sound, isn't it? The dolphins we're hearing now are all dead," says O'Barry. "Tomorrow there will be another group replacing them."[9] Elsewhere in the film, pointing to the cove and the smaller lagoon off to the side, O'Barry characterizes the place as "a dolphin's worst nightmare."

* * *

Writing about Freud's use of Tasso's romantic epic *Gerusalemme Liberata* to theorize persistent patterns of suffering in certain individuals, trauma studies scholar Cathy Caruth emphasizes that the experience Freud termed "traumatic neurosis" "emerges as the unwitting reenactment of an event that one simply cannot leave behind" (1996, 2). Tasso's hero Tancred unwittingly wounds his lover twice over: first, mortally, "in a duel while she is disguised in the armour of an enemy knight," Freud wrote. Then, "[a]fter her burial,"

> he makes his way into a strange magic forest which strikes the Crusaders' army with terror. He slashes with his sword at a tall tree; but blood streams from the cut and the voice of Clorinda, whose soul is imprisoned in the tree, is heard complaining that he has wounded his beloved once again.
>
> (Caruth 1996, 2; quoting Freud)

Caruth extends Freud's discussion of this traumatic repetition compulsion not only by developing the concept of traumatic "belatedness" (the notion that trauma is experienced belatedly as a haunting rather than at the time and place of "original" events) but also by emphasizing that the trauma that the wound "speaks" is *both* that of Tancred as (unwitting) perpetrator and that of Clorinda as the "other" who cries out. Caruth also meditates on the significance for psychoanalytic historiography of Freud's having turned to literature to explain his ideas.

If willing to psychologize Ric O'Barry, we might say that the distress discernible in his manner—evidence of his traumatization?—is produced from the inadvertent injury he himself inflicted on the dolphins by maintaining them in captivity, as well as from the recognition of the fatal plight of those whose cries he now hears:

> I watched them give birth. I nursed them back to health when they were sick. Had I known then what I know now I would have raised enough money to buy them away from the Seaquarium and set them free. That would have been the right thing to do. . . . But I was as ignorant as I could be for as long as I could be.

He reflects on a prior state of disavowal: he did not know (that he had wounded, was wounding, his beloved dolphins by keeping them captive at the end of the dock in the TV show's salt lake location) because he didn't *want* to know, while living the high life, buying a new Porsche each year. As Caruth writes,

> Just as Trancred does not hear the voice of Clorinda until the second wounding, so trauma is not locatable in the simply violent or original event in an individual's past, but rather in the way that its very unassimilated nature—the way it was precisely *not known* in the first instance—returns to haunt the survivor later on.
>
> (1996, 4)

But then O'Barry did come to consciousness, on the day that Kathy "cried out from the wound" (Caruth 1996, 3). He has spent the rest of his life to

date ("10 years building that the industry up" and "the last 35 years trying to tear it down") belatedly experiencing the trauma of captive dolphins
and bloody dolphin hunts all around the world and year after year in Taiji
(with his captor role now played by hunters); hearing in reality—and in his
head—the fishermen's metallic banging to build their wall of sound; and
hearing through recordings the desperation of the penned dolphins calling
to one another, to their separated babies or to humans in the vicinity.

Whether or not one gives credence to these speculations about O'Barry's
(or the dolphins') psychological state(s), the narrative structure of *The Cove*
may be said to parallel Tasso's story and that of traumatic experience in the
repetition of the slaughter and collective (in this case) human culpability.
The dolphins in the cove may well be construed as embodying "that other
voice" (Caruth 1996, 8)—acoustically present but linguistically incomprehensible to humans.

And yet, while recognizing *The Cove*'s major contribution to principled
advocacy for cetaceans and sustainable ecology, I would submit that it could
have moved more decidedly in the anthropomorphic direction it signals. Just
as Caruth regards the story of Tancred not only "as a parable of trauma and
of its uncanny repetition" but also as "aparable of psychoanalytic theory
itself as it listens to a voice that *it cannot fully know* but to which it nonetheless bears witness" (Caruth 1996, 9, emphasis added), *The Cove*'s spatial-
ized, traumatic witnessing is (*necessarily*, if we are of psychoanalytic bent)
deaf to certain voices and discourses to which it nevertheless bears witness.

Drawing on the Ancient Greek distinction between *zoē* ("expressing
the simple fact of living common to all living beings" [1]) and *bios* ("the
form or way of living proper to an individual or group" [1]) as well as the
Foucauldian concept of biopolitics, Giorgio Agamben suggests in *Homo
Sacer: Sovereign Power and Bare Life* that "[t]he fundamental categorical
pair of Western politics is not that of friend/enemy but that of bare life/
political existence, *zoē/bios*, exclusion/inclusion" (8) and perhaps also voice/
language, with these pairs coinciding in a "zone of irreducible indistinc-
tion" (9). The book's "protagonist" is a figure of Roman law included to be
excluded or existing to be "killed and yet not sacrificed" (8). But as Matthew
Chrulew has brilliantly analyzed, in a subsequent book Agamben extends
his discussion from the problem of "humanity animalised" to the consideration of non-human animal life itself (55). Chrulew explains that whereas
according to Agamben's broader argument "the exception of bare life is tied
to the attempt *to distinguish* humanity from animality," alternatively, in
Agamben's "historical task, man and animal are divided, and thereby *bound*
in the urgent repetition of that division, as a consequence of which both
animals, and animalised humans, are exposed to violence" (55, emphasis
added; Chrulew cites Agamben 2004). Approaching the film from the perspective of Agamben's and Chrulew's thought, we may come to understand
The Cove's dolphins as figures relegated *by the hunters* to the (low) status
of bare life—stripped of power, existing precisely for their "capacity to be

killed" (Agamben 2008, 8)—but held *by the film* to be exposed to violence in and through this "urgent repetition" of the catastrophic division of man and animal.

Perhaps, then, the protagonist- makers of the documentary observe their dolphin etiquette too well, filming without interfering, hanging back from ultimate recognition of dolphin existence. Biologists who came out publicly against a proposed International Whaling Commission (IWC) policy change that would have specified how many whales of each species could be "sustainably harvested" concluded instead "that maybe we shouldn't talk about *what* we're harvesting or harpooning, but *whom*" (Angier 2010, emphasis added).[10] The OPS crew have chosen to refrain from deep ecology's call for civil disobedience (Naess 29) or the type of forcible interventions for which environmental activist Paul Watson, founder of the Sea Shepherd Conservation Society, has become known; the film's first line includes Psihoyos saying, "We tried to do the story legally."

What I would suggest—with all due respect for the restraint, wisdom, legality and probable success over time of the route the film-makers took— is that their acceptance of the self-consciousness of dolphins complicates their choice not to intervene in the series of massacres to which the film bears witness. Agamben levels his thinking against this horrible exclusion that we must nevertheless acknowledge in order to change:

> ... until a completely new politics—that is, a politics no longer founded on the *exceptio* of bare life—is at hand, every theory and every praxis will remain imprisoned and immobile, and the [Aristotelian] "beautiful day" of life will be given citizenship only either through blood and death or in the perfect senselessness to which the [Debordian] society of the spectacle condemns it.
>
> (2008, 11)

Notwithstanding its insistence on the value of interspecies mutuality, in a context where humanitarian efforts are generally confined to humans, *The Cove* crashes up against an anthropomorphic conundrum.

WHALERS AS OTHERS

Of course, marine mammal hunters are also part of this ecosystem. If dolphins are acknowledged as sentient beings and some are even given credence as individuals, how are the "whalers" (Psihoyos's term) depicted? In a popular episode of the cartoon satire *South Park* (Parker and Stone 1997) entitled "Whale Whores" (Season 13, Episode 11), a horde of samurai warriors storm the Denver Aquarium during a child's birthday party at the Dolphin Encounter exhibit. We have just been introduced to the dolphins Trigger, Dolly and Bubbles, when, wielding their spears (and to the sound of someone crying "Oh, no, it's the Japanese"), the warriors proceed to stab the captive dolphins in the tank. With the bloodied carcasses in the foreground, and as the group runs off, one (cartoon) man and then another turns back towards the tank and yells in Japanese-accented English, "Fuck you, dolphin."

This brilliant and biting *South Park* satire captures the cultural and ethical complexities of the cetacean activism in which *Whale Wars* (an Animal Planet reality television series featuring Paul Watson in his role as captain aboard various ships seeking to stop Japanese whaling activities on the high seas) and *The Whale Warrior: Pirate for the Sea* (also with Paul Watson; directed by Ron Colby, 2010) as well as *The Cove* participate. Apart from rhyming with "whale wars," the exact significance of the title "whale whores" is unclear (nobody—nobody in the episode, that is—is literally prostituting him- or herself to satisfy an uncontrollable desire for whale meat—as in the urban expression "coke whore"). But it surely nails the anti-whaling caricature of "The Japanese" as addicted to going after whales (and small cetaceans, i.e., dolphins). Still, there is a rational explanation for this killing impulse, the *South Park* episode purports. From the prime minister on down, the Japanese mistakenly believe that a dolphin and a whale piloted the Enola Gay airplane that bombed Hiroshima. When they "find out" that the pilots were actually a chicken and a cow, the Japanese become "normal, like us" in their eating habits.

The Cove makes a point of distinguishing between (1) Taiji whalers and whale industry personnel and (2) Japanese populace and individual Japanese researchers and public authorities. The former are presented as willing to attack interlopers and greedy enough to poison their own children with mercury-laden dolphin meat in school lunches. At one point O'Barry is shown speaking through a translator to one of the hunters:

> O'Barry: Does he want to know if he's poisoning the bodies of other
>   Japanese that he's selling the meat to?
> Translator: No. He doesn't want to know.

At best, the hunters are misguided in their view, reported by Psihoyos, that mammals are gobbling up all the fish such that killing dolphins is "an issue of pest control." Or they are disingenuous. O'Barry says that fishermen have told him that dolphin hunting is their tradition. But he refutes the claim, noting that it cannot be a cultural tradition if the Japanese people do not even know that dolphin hunting is occurring—or that dolphin meat is being sold for food. Members of the public are presented either as victims of a media blackout of the issue or, in the cases of the researchers and officials, as having seen the light.

When we hear O'Barry say that "nobody has actually seen what takes place back there [in the killing lagoon] and so the way to stop it is to expose it," we are seeing footage of a fishing boat disappearing behind a rocky outcropping headed for the secret lagoon. Members of the Taiji Fishermen's Cooperative are willing to cull scores of dolphins for seaquariums, but they are not willing to harpoon thousands in public view. Thus, the circuit of secrecy and fear is closed. While the film-makers attempt to infiltrate the cove without being seen, the fishermen also try to avoid being immediately observed or filmed.

The latter are not completely successful. Following O'Barry's statement that the cove killing action is heretofore unwitnessed, there is a sequence of film-makers versus fishermen: shaky, angled shots, pushed off balance by the jockeying for position; one guy holds up a "Don't Take Photos!" sign to block the camera's view while another swats the lens with his cap.

Of course, images from the planted cameras are indeed retrieved. They show hunters pulling their boats alongside the cordoned off dolphins, killing dozens of animals with multiple stabs, and grappling them into the boats with barbed hooks. One key camera has been positioned to aim outward from shore, another aims shoreward from the mouth of the cove. Different angles and shot distances have been achieved by varying the zoom of the cameras on each of multiple nights of filming, and through digital processing.

In the notorious killing sequence near the end of the film, we see the water in the cove turned blood red, in strong colour contrast to the white and aquamarine painted boats that glide along its surface. A scuba diver surfaces, blows water from his snorkel, and dives back down in search of dolphin carcasses on the bottom (where visibility must be next to nil), his bright yellow flippers the last we see as he is enveloped by the opaque fluid. The soundtrack is a mix of dolphin whistles recorded under water, the above-water ambient sounds of the hunters and the musical score. An improvised flute melody is heard, the wind instrument invoking and harmonizing with the voices of the dolphins.

This is the material (snippets of which are included in the film's trailer) that has kept audiences away while at the same time guaranteeing the film's terrible legitimacy as a document of what has occurred and continues to this day: the low-tech but highly organized, profitable and prolific annual killing of tens of thousands of socially attuned and self-conscious creatures.

I would maintain that the dolphin slaughter footage is shocking because the actions documented are understood by those in the film and many in the film's audiences as cruel and unethical, and by Psihoyos himself as genocidal, for being directed at a particular group it is assumed may be decimated with impunity. Whereas the Nazis succeeded in precluding films of killing during the Holocaust such that "there is only one known piece of motion picture footage [showing killing action], lasting about two minutes" (Hirsch 1),[11] the fishermen in Taiji were not as successful.

Psihoyos's specific reference is to the "banality of genocide," which phrase he uses (on the commentary track) to describe the casual attitude of the whalers in the killing sequence[12] (see Figure 9.1). In this regard, and illustrative of the incredible revelations clandestine filming may produce, we eavesdrop on a conversation among the whalers at dawn preparing for the day's labours. Standing around the fire, reminiscing about past whaling experiences, the cohort of Taiji hunters display a casual camaraderie. A few minutes later, once we have seen the killing action, a shot is inserted of one of the men pouring blood-infused water to douse the fire. The whalers might be just any group of working men, discussing the physical rigours of the job in a matter-of-fact manner, smoking as they chat amongst themselves, unwinding as they bring their catch to shore. But this is big business. A graphic title has informed us previously that an animal for a dolphinaria might garner $150,000 and the carcasses add up to hundreds of thousands of pounds for sale.

Should a question arise at to the ethics of filming these men without their knowledge, one might address if not resolve it with reference to Claude Lanzmann's secret filming of Franz Suchomel, SS Untersharführer in *Shoah* (Lanzmann 1985). There we see and hear Lanzmann promise Suchomel that his name won't be used (not to mention his image, which Suchomel does not know is being captured). But he is interviewed, named and, therefore, singled out by the film as a Treblinka death camp participant.[13] The whalers, on the other hand, are not singled out for individual criticism (nor have any promises been made them).

The exception is Private Space. He has been given this moniker by the film-makers, and an off-screen voice offers the opinion that these are the only words he knows in English. We see him throughout the film, including in the killing sequence where he jumps out of a boat and walks up the beach to warm his hands over the fire pit. Often we see him interacting with the

Figure 9.1 Dawn in the killing lagoon. Image from *The Cove* courtesy of Oceanic Preservation Society.

film-makers: squatting by the side of the road to accost them as they come along, shouting into the lens—exerting a territorial imperative that must be defended through repeated engagements with the film-makers who did finally succeed in infiltrating "Private Space's private space" (Psihoyos, commentary track). There is also a blurry image of Private Space looking right into the lens.

By filming and including in the finished work their own interactions with Private Space and others who would defend *The Cove*'s secret, the film-makers foreground their "interactivity": "a sense of . . . *situated* presence and *local* knowledge that derives from the actual encounter of film-maker and other" (Nichols 44, original emphasis). The fishermen's combative and self-serving attitudes are exposed.[14]

Are the Japanese in *The Cove* "whale whores"? Not in the film's own terms. It is only this particular cohort of Taiji whalers and those who maintain them in business whose behaviour is impugned by the film. In fact, two conscientious Taiji town councilmen have made themselves available to be interviewed. "If dolphin meat is used for school lunch it brings about terrible results," states one of the men.[15] The film's distinction among various constituencies is to be applauded, I believe, especially since the number of dolphin hunters involved in the slaughter is known to be small in relation to the overall number of fishermen in the area: perhaps something like two or three dozen dolphin hunters among a population of five hundred fishermen (McCurry 2009, 2010).

MAPPING *THE COVE*

What does it mean to speak—and listen—from the spot itself, from the very place where violent, catastrophic events occurred and continue to occur on an ongoing basis? *The Cove*'s physical geography is a key factor in its use as a net and buoy dolphin death camp. Dolphins really were and, as I write this, *are* being harpooned and grappled there. Through the film's "site-seeing," in Giuliana Bruno's term, and site-hearing (less alliteratively) converged on Taiji, we experience vicariously the affective witnessing-from-the-place in which O'Barry, Psihoyos, Cruickshank and the others participate.

And yet, moving into position with trauma studies scholar Cathy Caruth, we may recognize as well that place is never firmly located, and there is an outside. Alluding to the spatial dimension of traumatic experience, Caruth writes that "[T]he impact of a traumatic event lies precisely in its belatedness, *in its refusal to be simply located*, in its appearance *outside the boundaries of any single place* or time" (1995, 9, emphasis added).

Further inspired by the insights of critical human geography and choosing at this point in the article to read expansively, I understand space in *The Cove* as *both* heavily material and also profoundly unassimilable and traumatic. "What is geography beyond the charting of land masses, climate zones, elevations, bodies of water, populated terrains, nation states, geological strata and natural resource deposits?" asks critical human geography-influenced art historian Irit Rogoff (21). She provides a response, writing, in part, that geography is a "mode of location" and "epistemic category . . . grounded in issues of positionality" (21). Accordingly, I would submit that along with the cove's existence as a physical feature goes its meaning as an epistemological regime and site of discursive contestation.

Whereas O'Barry vows to concentrate on the "one little body of water" as if its boundaries were clear, the film itself presents the cove as subject to multiple, competing constructions. At one point OPS member and photographer/cinematographer Charles Hambleton describes how Japanese officials presented the film-makers with a map indicating the areas that were off limits, and how he asked to keep the map—ostensibly to know where *not* to go, but actually to reference the exact spots to penetrate and reconstruct in the filmic idiom. The cove with its cliffs, mountains and engineered tunnels, in addition to being presented through garnered documentary footage, is also re-presented as a digital animation created at significant expense. These various maps do more than merely orient the viewer to the geography at hand. They also serve, first of all, to facilitate the filmic occupation of a furiously contested space, second, as a reminder that mapping is not an ideologically neutral proposition, and third, as a metonym for the film's work of constituting and reconstituting—the "fatal environment"[16] it might seem only to approach, observe and record. In its audio-visual reconstruction, the cove becomes the purview of film-makers and spectators who together infiltrate, inhabit, survey, map, shoot, record, fragment, consider and ultimately mosaic back together—*in a different form*, precisely, as *The Cove*—this would-be off-limits space (Figure 9.2).

In and through our experience of *The Cove*, we may come to know more deeply the central insight of Henri Lefebvre's magisterial work, *The Production of Space* (Lefebvre 229) that space itself is a social phenomenon and that "occupied space gives direct expression . . . to the relationships upon which social organization is founded."

In addition to construing the lagoon as both material and multiply produced, we may also observe the film's rendering of this space as both an exceptional "geography of atrocity" (Robert Jan van Pelt uses this term in relation to Auschwitz-Birkenau during his participation in Errol Morris's film *Mr. Death: The Rise and Fall of Fred A. Leuchter, Jr.*, 1999) and also, somewhat paradoxically, as the hub of a global network. The dolphin trade practices, permitted by the Japanese central government and emanating from the killing cove, are exceptional not only in their robustness that the film amply displays, but also in comparison with those of most western dolphinaria. These latter rely on breeding programmes to replenish their Bottlenose dolphin population even though trade in multiple dolphin species is regulated through the International Trade in Endangered Species of Wild

Figure 9.2   Two of the clandestine camera positions, approximate reverse angles, from which the space of the cove is (re)constituted by the film-makers. Mapped by Janet Walker in Google Earth.

Fauna and Flora (CITES) and Bottlenose are not endangered (Alabaster 2010). But the film also demonstrates Taiji's centrality and linkages to the seaquarium industry's "dissipated accountability," to adopt Anil Narine's significant concept.[17] Passages in the film introduce the International Whaling Commission, as, precisely, an international body, albeit one that Japan attempts mightily to influence.[18] Hence the logic of the bright red curved lines that we see superimposed over a map of the world radiating outward from Taiji to points elsewhere, presumably designating dolphinaria, including in North America.

Even more crucial for this consideration of the extent and limits of the film's sea life/sea food cartography is the question of how it figures the wider problem of wild fish depletion. A time-lapse sequence of inventory movement at Tokyo's Tsukiji Market, the biggest wholesale fish market in the world, is interrupted by a title card that reads, "A 2006 report in the journal *Science* predicts a total collapse of the world's fish stocks within 40 years at the current rate of fishing." As *Cove* interviewee Michael Illiff of the Institute of Antarctic and South Ocean Studies at the University of Tasmania states, "They [the Japanese] have a real fear that they'll run out of food. What more logical thing can they do but to catch whales to replace them?" This latter sequence, brief though it may be, does serve to locate dolphin hunting within a broad environmental and "eco-economic" (Cubitt 2005) network of food production. But overfishing is neither extensively explored nor graphically prominent in the film. A new cartography might therefore be envisioned, with fish *and* non-human *and* human mammals included, and the small body of water off the Taiji coast as but one node in a mesh network with multiple intersections and lines radiating both in and out.

Imagining the contours of this broader ecosystem also enables as a rather different view of Taiji-based fishermen. The hidden camera and microphone that captured the hunter's apparently calloused attitudes around their campfire also captured the following oral historical account of marine life depletion:

> In Midway, Hawaii, you know Midway, Hawaii, I saw sperm whales from horizon to horizon. Just like dolphins. There was a time when sperm whales were as plentiful as dolphins. When I was in Chile I saw blue whales from horizon to horizon. Wherever you looked, the ocean was truly black. It was covered with blue whales. My arms were exhausted.

However inadvertently and indirectly, the sequence suggests (and Simon Hutchins has confirmed) that the older men among the Taiji group are former off-shore whalers, once employed on distant water vessels and now working, by necessity, in coastal waters.[19] Thus, from the film itself one may glean that the lives of the whalers being depicted have been negatively affected by unsustainable fishing practices.

Researching beyond the film, international studies such as those conducted by the UN Food and Agriculture Organization (FAO) (Pulvenis de Séligny et al. 2008) and UN Environmental Programme (UNEP) (2004) and a 2010 study by Canadian, US and Australian researchers (Swartz et al. 2010) report the dire depletion of world fish stocks and a declining global seafood catch.

The Japanese fishing industry, among the largest in the world, has been greatly impacted by this situation. Commercial fleets have long plied territorial and distant waters. But whereas in the 1950s, fishing's ecological footprint was concentrated in the coastal waters off Europe, North America and Japan, by 2005 it had expanded into the Southern Hemisphere and the high seas as new waters had to be sought for commercial exploitation (Swartz et al. 2). A recent report on Pacific Island Fisheries notes the "steadiness in the Japanese purse-seine [a type of net fishing] fleet" in the area (Gillett and Cartwright 9). Even so, the Japanese fishing industry is declining due to a lack of wild fish and foreign competition that has lowered the price of the resource. Since the industry peak in the 1980s (Swartz et al. 2), the number of Japanese fishing boats has dropped (Pulvenis de Sélignyet al. 46), as has the total deep-sea catch by three quarters between 1970 and 1999 (Kristof A4). The number of people employed in fishing has fallen correspondingly due to these and other factors, including the employment of lower paid foreign workers on what remains of Japan's deep-sea fishing fleet (Pulvenis de Séligny et al. 43).

With fish stocks in coastal waters severely exploited by commercial fleets, many fishing villages have lost population or their fishers have been forced to seek other work (Fackler 2008). Nicholas Kristof's (1999) *New York Times* commentary, "Ah, When Nets Were Full, and So Was Life at Sea," sketches a very different portrait of the fishing community than does *The Cove*. Kristof reports that "nearly half of Japan's 290,000 remaining fishermen are in their 60s and 70s" and the younger generation cannot make a living from fishing. Citing the statistic that "[o]nly 205,000 Japanese households—one-half of 1 percent of the total—are now engaged in fishing," he proceeds to describe the changed circumstances of coastal fishermen:

> [T]hey get most of their income from other sources. Some run sports fishing charter boats or even operate sightseeing boats for whale-watchers from the cities. "There are just fewer and fewer fish out there," said Tanisaku Makiyama, a 44-year-old fisherman. "Many of the young fishermen are still single, unable to find a wife, because their incomes are so low."
>
> (A4)

This description likely corresponds to the plight of the minority of Taiji fishermen who engage in dolphin hunting, among other workingmen whose occupations have changed drastically with the situation. According to Simon Hutchins, the younger fishermen who appear in the film hunt dolphins during half the year and fish in the alternate season, thus expanding their options. Large commercial fishing fleets—and not these dolphin hunters—have depleted whale and fish populations, and thrown their former employees out of work through short-term, profit-driven and unsustainable practices. The whalers' memories of work-produced exhaustion harken back to years of fuller employment. We may choose to compare these fishermen to workers at a Hormel plant or other slaughterhouses in the US meat and poultry industries whose livelihoods have been compromised by the consolidation of wealth among the executives and stockholders. If only "the Japanese" ate "chicken" and "cow"—as well they may one day if aquaculture proves insufficient to replace wild fish.

The dolphin hunters have their own video cameras. I wish I could see the footage they obtained: the reverse perspective on this contested territory. Singling out whalers as the villains of the piece is problematic to the extent that it inhibits an ecosystemic view not only of the multi-billion dollar marine mammal corporate infrastracture around the world, but also of the ecological catastrophe of worldwide wild sea life depletion.

Singling out dolphins is also problematic. Although I have swum with the film's anthropmorphism, the philosophy is limiting, especially where it blurs into anthropocentrism. The film to some extent prizes dolphins because they are sentient, like us and not like fish, rather than because we and all the other animals are part of a larger ecosystem.[20] As discussed, one implication of this filmic logic is that *not* intervening in the immediate massacres to which the film bears witness could be construed as unconscionable. Perhaps even more saliently, the film's anthropocentrizing anthropomorphism inhibits acknowledgement of an ecocentric perspective in which sustainable hunting and fishing practices are seen as imperative, not only for the sake of dolphins and coastal Japanese, but because "the entire ecosphere and ecological systems are thought to be of value" (Katz et al. xiii). Why not understand the dolphins in the cove as embodying not just "that other voice" (Caruth 1996, 8)—acoustically present but linguistically incomprehensible to humans—but as one among many other voices to which we humans should attend?

EAVESDROPPING AS WITNESSING

The passages in the film showing dolphins in the wild are a huge relief, although we watch them, from the start, with prior and *Cove*-conveyed knowledge of dolphin captivity and slaughter. Distributed throughout the film, these sea-blue sequences of dolphins swimming and jumping above and surfing inside of waves are made even more expressive by the accompanying score featuring the music sounding to my ear like a plucked harp. In one such sequence, along with music and images of dolphins, we hear and see surfer and cetacean activist Dave Rastovich express what it feels like to look down into the water while surfing and see dolphins alongside as if in a glass case, and how his life was once saved by a dolphin who t-boned a shark coming right at him. In the main sequence of this nature—and it is truly transporting—free-diver Mandy-Rae Cruickshank is photographed underwater swimming with wild dolphins. We hear her narration—a rare human female voice in the film—intercut with an interview in which her eyes glow with excitement as she describes the distinct feeling of "understanding" that has passed between her and a dolphin on such an occasion.

> When you're out swimming in the ocean and you have whales and dolphins come by you, it is one of the most incredible experiences ever. It's so humbling that this wild creature would come up and be so interested in you. It's unbelievable really. Even though there's obviously no words spoken, you really feel like you're on some level communicating with them. Like there's an understanding between the two of you.

"I don't normally touch anything in the water," she continues, in keeping with Dudzinski's and Frohoff's eavesdropping etiquette. But in this case, the dolphin of its own volition swam right up and rolled into her hand, so she rubbed its belly. With Cruickshank and the film, "we" submerge to be with the dolphins. Our diving down constitutes an action comparable if opposite to their spy-hopping or looking from below to above the surface; a kind of reciprocal breaching and perspective. Here there is no need to be a clandestine or even "boring" observer. Zoomorphically speaking, the dolphins seem to welcome the underwater visits of the humans and the resulting engagement in their water world.

In a way, we, as viewers and auditors of the film, do come to listen after all. With reference to trauma theory, we may comprehend the address of the voice in *The Cove*, and the ecological crisis that the mercury poisoning (including of dolphins themselves and not just "their meat") and decimation of dolphins represents, "not as the story of the individual in relation to events of his own past, but as the story of the way in which one's own trauma is tied up with the trauma of another" (Caruth 1996, 8). This form of listening is also tantamount to "bearing witness" as defined by clinical psychiatrist and theorist of Holocaust-related trauma, Dori Laub. It is a form of listening "to an event that has not yet come into existence, in spite of the overwhelming and compelling nature of the reality of its occurrence" (Laub 57). It "includes its hearer, who is, so to speak, the blank screen on which the event comes to be inscribed for the first time" (Laub 57), and "may lead, therefore, to the encounter with another, through the very possibility and surprise of listening to another's wound" (Caruth 1996, 8).

Imagine if dolphins had their own cameras (even though, as Dr John Potter remarks in the film, they don't have hands). *National Geographic* is engaged with scientists in an ongoing project to attach sensing devices including cameras to marine creatures. "By allowing us this animal's-eye view, Crittercams help to solve scientific mysteries," the Crittercam Chronicles website reports. In one such video posted to YouTube, we see footage from a small camera suctioned to the back of a pilot whale showing whales "socializing at depth," according to the narration, and diving down to feed at the astonishingly low depth of 2,300 feet.[21] Of course, this project was not initiated nor carried out by marine mammals. But it may inspire thinking about life and death in the cove from a dolphin's-eye perspective.

Another way to infiltrate the cove would have been to attach cameras to the dolphins out at sea before they were driven in. This procedure might have garnered individualized footage of an animal under attack, but of course the cameras would have been difficult if not impossible to retrieve. In any case, I wish the dolphins would learn to give the Taiji cove a wide berth and to warn one another.[22] This may not be entirely far-fetched. Interviewee and film-maker–activist Hardy Jones mentions another spot, Iki Island, where dolphins once were hunted, but are no longer found. Rather than assuming the pods that passed through the area have been wiped out, I prefer to imagine that they have learned to avoid the spot.

SURFACING[23]

During its last few minutes, the film draws viewers' attention to its production process and image activism, through the incorporation of three "reflexive" documentary moments (see Nichols 1991). In the first, we see from over his shoulder as Psihoyos holds up to the face of Hideki Moro-nuki (Deputy Director of the Fisheries Agency of Japan) an early iPhone on which the footage Psihoyos reports having just seen is being played. Only the back of the iphone (and therefore not its screen) is visible in the shot, such that audibility is prioritized: those "eerie" dolphin vocalizations heard previously in the killing sequence. "When and where did you take this"? Moronuki asks. In the second reflexive instance, we follow Ric O'Barry as he bursts into a meeting of the International Whaling Commission with a monitor strapped to his torso. This time we can discern, even at a distance as O'Barry turns to face the room and the news cameramen covering the IWC meeting rush-in, that the footage playing on the monitor is that of the dolphin slaughter we have seen just previously in the film—gliding tur-quoise fishing boats and the inlet running red with blood. The third instance ends the film: Ric O'Barry, again wearing the monitor harness, standing on a street in Tokyo exposing images of dolphin killing to the pedestrian traffic. The time-lapse photography with which this three-camera sequence is shot makes geometric patterns of the umbrella-carrying passers-by, and of the few who stop. It suggests the long duration of O'Barry's staunch commitment.

These moments of forcible exposure to watchers with the film reflect and affect our own experience as audience members, including the hesitation many felt to see the film after the blood of the trailer. Here we are given to understand that the film text consists, reflexively, of delivering and receiving the story as well as investigating and constructing it. Just as the network of profitability for whaling extends around the globe, so too the film's life as a media object extends from below the surface of the cove to the overland routes of its film prints and the fibre-optic paths of its cable-casting.

As I drafted this chapter, protestors in Japan were continuing to press the boycott of theatrical screenings, but *The Cove* had just debuted on US television (on 27 August 2010) in conjunction with the Animal Planet series *Blood Dolphin* (2010) designed to update and extend knowledge of the fraught relationship between humans and dolphins.

Still thinking about that one little body of water, I navigate my way to the cove. Unable to make the trip in person, I travel via the geographic information programme Google Earth. Although resolution is lacking as I zoom down into the cove, a click on the icon of a local hotel brings up photos of the gorgeous, tranquil surroundings and also an enticing whale-watching video. This is the view of Taiji *The Cove* deems violently deceptive but which O'Barry and company must now help cultivate as a financial alternative to whaling (McCurry 2009[24]). In the meantime, somebody has added a photo of bloodied dolphin carcasses labelled "that's so sad" to mark the killing cove.

By means of amphibious locomotion and psychic trauma, *The Cove* encourages audiences to imagine how the world would be different if we all eavesdropped on dolphins underwater instead of hunting them. Conceptualized as a device for mapping and navigating the ecology of the dolphin-human interaction, the film furthers as I have argued a pooching, spherical optic and echolocation of life and death matters in and beyond the killing lagoon. *The Cove* has inspired me—and I hope us—to think ecosystemically about our fellow creatures under and above the waters of this liquid planet.

ACKNOWLEDGEMENTS

I am grateful to Anil Narine for having inspired this chapter through his invitation and concept of "eco-trauma." My thanks go also to Simon Hutchins, *The Cove*'s expedition and technical director, for his impressive presentation on our campus and for his time, knowledge and consideration in answering my follow-up questions via e-mail while out at sea. The comments and work by my students in the course "Films of the Natural and Human Environment" (Fall 2009, 2010, 2011, 2013), especially Claire Johnson and Lindsey Parker, enlivened my thinking. Lisa Parks and Chuck Wolfe encouraged me to write about this film, and Jaimie Baron's "transgressive documentary" panel call, Nicole Starosielski's work on under water media and Ariel Nelson's comments from the perspective of transitional justice were fundamental. Finally, I am grateful to Deane Williams, Sally Wilson, Tim Mitchell and the anonymous reviewers of this essay for its initial appearance in *Studies in Documentary Film* 7:3 (2013), and for their permission to reprint the essay as a chapter in this book.

NOTES

1. Dolphins are not whales, but they are classified along with porpoises and whales within the order cetacea. According to Save Japan Dolphins, an Earth Institute Project, the Japanese government Fisheries Department issues annual permits to kill dolphins, porpoises and small whales. This source places the number of permits issued in 2011 at 19,300. http://savejapandolphins.org/take-action/frequently-asked-questions, accessed 21 June 2012.
2. One event that occasioned such debate was a panel discussion I convened with presenters Toni Frohoff, Simon Hutchins and Peggy Oki. "*The Cove*: Thinking Through the Dolphin-Human Interaction," University of California, Santa Barbara, 4 November 2009. I would also like to cite, with gratitude, the lively discussion that followed my presentation of "Eavesdropping in *The Cove*: Interspecies Ethics, Public and Private Space, and Media under Water" as a conference paper at Visible Evidence 18, New York, 11–14 August 2011.
3. Jaimie Baron designed and chaired the Visible Evidence panel "Transgressive Bodies, Bodily Transgressions: Exposure and Occlusion in Recent Documentary Films," on which an early version of this chapter was presented as a talk.
4. Multiple species of dolphins are targeted off the coast of Japan. See http://savejapandolphins.org/take-action/frequently-asked-questions, accessed 20 June 2012.

5. In point of fact, although *The Cove* contrasts *Flipper*'s anthropomorphic deception with its own avowed pro-dolphin sensibility, we may discern in individual episodes of the *Flipper* TV show a critique of animal captivity, if not anthropomorphism per se. For example, Season 1, Episode 6 (24 October 1964), is entitled "Dolphin for Sale" and concerns a dishonest fisherman who "lures Flipper from the preserve and wants to sell him to a circus" (Wikipedia entry for "Flipper [1964 TV Series]"). Later that season ("Mr. Marvello," Episode 9, 7 November 1964), a ventriloquist tries to acquire Flipper for his circus act by convincing kids and others that Flipper talks. This deceit is characterized as such and the plot itself promotes Flipper's continued "freedom" in the lagoon.

6. Hutchins reports that there were at least six equipment planting and retrieving missions and two missions to plant hydrophones (Hutchins 2010).

7. Cameras were dubbed "Nest" (looks like a bird's nest), "Thermal" (used during the secret missions to detect the presence of whalers, but ended up documenting OPS's own activities), "Rocks Cams," "Heli Cam" and "Blood Cam." There was also an unmanned drone (painted to look like a whale and named Kathy) created for the purpose of aerial photography because, as indicated in the DVD special features, "even if the blimp didn't succeed and we got caught . . ., everybody loves a balloon".

8. Whale vocalizations are also used as a sound bridge to the historical footage of a 1971 Save the Whale movement rally in Trafalgar Square, London. There, the whale sounds—haunting cries—recorded by pioneering cetacean researcher Roger Payne, Ph.D., were amplified and broadcast over the gathered crowd. *Cove* interviewee Dr John Potter, underwater acoustics consultant, remarks on the sorely ironic one-way communication effect of using sign language to train and command captive dolphins, since, as he quips, "dolphins don't *have* hands."

9. The film also includes a passage of Cruickshank on the shore of the cove crying as she watches a wounded baby dolphin that had somehow leapt one of the barrier nets take its last breath and sink below the surface. In voice-over narration intercut with on-camera testimony, she describes in tearful words and graceful hand gestures the death of this dolphin separated from its pod, which had been driven into the killing lagoon.

10. Returning from this perspective to the mirror test for dolphin self-recognition, I wonder whether a given dolphin subject might frame the problem of being marked with a pen as one of "What has this human done do me"?

11. Joshua Hirsch reports that the existing film is 8mm amateur footage of Jews who had been rounded up and shot by a firing squad in a pit in Latvia, and that the film is held at the Yad Vashem Holocaust Memorial Museum in Jerusalem with a copy at the United States Holocaust Museum in Washington, DC.

12. Cavalieri explicitly discusses whaling in the context of genocide.

13. I have always been fascinated by the cock-eyed honesty of Lanzmann's having left in the film the promise to Suchomel that he makes in the very act of breaking it. He could have abandoned the material on the cutting room floor along with shots of the van from which the filming was handled remotely. Still, audiences would have guessed from the quality of the footage that it was filmed with a hidden camera, even without Lanzmann's on-screen acknowledgement of his subterfuge (see Erens 1986).

14. These interactions have much in common with what one could term the "negative" interactivity Michael Moore purposely features when building a film around his failed attempts to get in to see General Motors CEO Roger Smith (*Roger & Me*, 1989) or to get a K-Mart employee to remove boxes of bullets from the store's shelves (*Bowling for Columbine*, 2002). But

there are "positive" interactions as well, both in Moore's *oeuvre* and in *The Cove*, when the film-makers and interview subjects are in agreement and talk together about matters of mutual concern (see footnote number 15).

15. Through an end title card, we learn that the councilmen did manage to have the dolphin meat removed from Taiji school children's lunches. Psihoyos also informs us in voice-over that they risked "if not their lives, then their livelihoods" to speak out. And on the commentary track he tells us that, subsequent to the filming, these officials were ostracized by the community and forced to reject the film. One man had to flee the town with his family because it became too dangerous to remain.
16. This term is adapted from Slotkin.
17. My thanks to Anil Narine for suggesting this concept and phrase drawn from his own research (Narine 2010a, 2010b).
18. The film depicts, for example, the collusion between Japan and certain Caribbean member states that sell their votes to the wealthier nation for material gain.
19. "The Old Timers were certainly experienced off shore whalers," he wrote (Hutchins 2010).
20. O'Barry takes care in certain moments of the interview with him to emphasize dolphin "sentience" over "intelligence," presumably to avoided the very homology with the human brain that the film does nevertheless suggest.
21. www.youtube.com/watch?v=F8kEJyur_C0, accessed 21 June 2011.
22. Cetacean volition is discussed by Paola Cavalieri, who begins her essay "Cetaceans: From Bare Life to Nonhuman Others" with an account of a white sperm whale who "terrorized whalers," it was thought deliberately, off the coast of Chile in the first half of the nineteenth century.
23. My use of this term is inspired by Nicole Starosielski and her brilliant work on "media under water."
24. McCurry quotes O'Barry as follows:

Japanese people have to get involved in this issue. There are groups out there calling for a boycott of Japanese goods, but I am involved in an anti-boycott campaign. We want people to go to Taiji and spend money in its hotels, restaurants and shops. We want to stimulate Taiji's economy, not ruin it.

(2010)

WORKS CITED

Agamben, Giorgio. *Homo Sacer: Sovereign Power and Bare Life*. Trans. Daniel Heller-Roazen. Palo Alto: Stanford University Press, 2008. Print.
——*The Open: Man and Animal*. Trans. Kevin Attell. Stanford: Stanford University Press, 2004. Print.
Alabaster, Jay. "'Cove'" Oscar Won't End Taiji Dolphin Kill," *The Japan Times*, 10 March 2010. www.japantimes.co.jp/news/2010/03/10/national/cove-oscar-wont-end-taiji-dolphin-kill/. Accessed 10 December 2010. Web.
Angier, Natalie. "Save a Whale, Save a Soul, Goes the Cry," *New York Times*, 25 June 2010. www.nytimes.com/2010/06/27/weekinreview/27angier.html?scp=1&sq=angier,%20natalie&st=cse. Accessed 12 July 2010. Web.
Animal Planet. *Blood Dolphins*, USA: Animal Planet. Aired on Friday nights beginning 3 September 2010. TV Program.
Bowling for Columbine. Dir. Michael Moore. Alliance Atlantis Communications, 2002. Film.
Bruno, Giuliana. *Atlas of Emotion: Journeys in Art, Architecture, and Film*. New York: Verso, 2002. Print.

Caruth, Cathy. Ed. *Trauma. Explorations in Memory*. Baltimore: The Johns Hopkins University Press, 1995. Print.

———. *Unclaimed Experience: Trauma, Narrative, History*, Baltimore: The Johns Hopkins University Press, 1996. Print.

Cavalieri, Paola. "Cetaceans: From Bare Life to Nonhuman Others," *Logos: A Journal of Modern Society & Culture*10:1 (2011): n.p. www.logosjournal.com/cetaceans-bare-life-nonhuman-others.php. Accessed November 2013. Web.

Chrulew, Matthew. "Animals in Biopolitical Theory: Between Agamben and Negri," *New Formations*76, Winter (2012): 53–67. Print.

The Cove. Dir. Louie Psihoyos. USA: Cos. Diamond Docs, Fish Films, Oceanic Preservation Society, Quickfire Films. Distribs. Lionsgate, Roadside Attractions, and international distributors, 2009. Film.

Cubitt, Sean. *EcoMedia*, Amsterdam and New York: Editions Rodopi B.V., 2005. Print.

Dudzinski, Kathleen M. and Toni Frohoff. *Dolphin Mysteries: Unlocking the Secrets of Communication*. New Haven: Yale University Press, 2008. Print.

Erens, Patricia. "*Shoah*," *Film Quarterly* 39:4, Summer (1986): 28–31. Print.

Fackler, Martin. "Japan Tests Technology to Lift Fishing Industry," *New York Times*, 26 December 2008, p. B3. Print.

Flipper. Created by Jack Cowden and Ricou Browning.18 September 1964–15 April 1967. Los Angeles: NBC TV, 1964–1967. TV Program.

Freud, Sigmund. *Beyond the Pleasure Principle: The Standard Edition of the Complete Psychological Works of Sigmund Freud*. Trans. James Strachey, Anna Freud, Alix Strachey and Alan Tyson. 24 Vols, Vol. 18, Chapter 3, London: Hogarth, 1953–1974. Print.

Frohoff, Toni. Panel presentation, University of California, Santa Barbara, 4 November 2009. Panel.

Frohoff, Toni, Simon Hutchins, Peggy Oki and Janet Walker. "*The Cove*: Thinking Through the Dolphin-Human Interaction," Panel convened by Janet Walker, University of California, Santa Barbara, 4 November 2009. Panel.

Gillett, Robert and Ian Cartwright. *The Future of Pacific Island Fisheries*. New Caledonia: Secretariat of the Pacific Community (SPC), Noumea Cedex, 2010. Print.

Hirsch, Joshua. *Afterimage: Film, Trauma, and the Holocaust*, Philadelphia: Temple University Press, 2004. Print.

Hutchins, Simon. Panel presentation, University of California, Santa Barbara, 4 November 2009. Print.

——— Personal e-mail communication, 19 November 2010. Web.

Katz, Eric, Andrew Light, and David Rothenberg. "Introduction: Deep Ecology as Philosophy." *Beneath the Surface: Critical Essays in the Philosophy of Deep Ecology*. Ed. Eric Katz, Andrew Light, and David Rothenberg. pp. ix–xxiv. Cambridge, MA: The MIT Press, 2000. Print.

Kristof, Nicholas. "Ah, When Nets Were Full, and So Was Life at Sea," *New York Times*, 9 July 1999. www.nytimes.com/1999/07/09/world/oshima-journal-ah-when-nets-were-full-and-so-was-life-at-sea.html?pagewanted=2&src=pm. Accessed 10 December 2010. Web.

Laub, Dori. "Bearing Witness, or the Vicissitudes of Listening." *Testimony: Crises of Witnessing in Literature, Psychoanalysis, and History*. Ed. Shoshana Felman and Dori Laub. pp. 57–74. New York: Routledge, 1991. Print.

Leake, Jonathan. "Scientists Say Dolphins Should Be Treated as Non-Human Persons," *New York Times*, 3 January 2010. www.timesonline.co.uk/tol/news/science/article6973994.ece. Accessed 21 July 2010. Web.

Lefebvre, Henri. *The Production of Space*. Trans. Donald Nicholson-Smith. Malden, MA: Blackwell Publishing, [1974] 1991. Print.

Mitchell, Sandra D. "Anthropomorphism and Cross-Species Modeling." *Thinking with Animals: New Perspectives on Anthropomorphism.* Ed. Lorraine Daston and Gregg Mitman. pp. 100–17. New York: Columbia University Press, 2005. Print.

McCurry, Justin. "It's Dante's Inferno for Dolphins," *GlobalPost*, 25 September 2009. www.globalpost.com/dispatch/japan/090924/its-dantes-inferno-dolphins?page=0,1. Accessed 21 November 2013. Web.

———— "Japan's Dolphin Slaughter: Cruelty or Custom," *GlobalPost*, 30 September 2010. www.globalpost.com/dispatch/japan/100929/taiji-dolphin-slaughter-cove. Accessed 21 November 2013. Web.

Le Monde du silence/The Silent World. Dirs. Jacques-Yves Cousteau and Louis Malle. France: FSJYC Production, Requins Associés, SocietéFilmad, Titanus, 1956. Film.

Mr. Death: The Rise and Fall of Fred A. Leuchter, Jr. Dir. Errol Morris. UK/USA: Channel Four Films, Fourth Floor Pictures, Independent Film Channel, Scout Productions, 1999. Film.

Naess, Arne. *Ecology, Community and Lifestyle.* Trans. and Ed. David Rothenberg. Cambridge: Cambridge University Press, 1989. Print.

Narine, Anil. "The Cinematic Network Society: Ethical Confrontations with New Proximities to Human Suffering in the Information Age," Doctoral dissertation, School of Communication, Simon Fraser University, British Columbia, Canada, 2010a. http://summit.sfu.ca/item/11840. Print.

———— "Global Trauma and the Cinematic Network Society," *Critical Studies in Media Communication*, 27:3 (2010b): 209–32. Print.

National Geographic. *Crittercam Chronicles.* National Geographic, 1996–2013. www.nationalgeographic.com/crittercam/. Accessed throughout July 2010. Web.

Nichols, Bill. *Representing Reality: Issues and Concepts in Documentary.* Bloomington and Indianapolis: Indiana University Press, 1991. Print.

Oceanic Preservation Society. www.opsociety.org/. Accessed throughout July 2010. Web.

Pulvenis de Séligny, J.-F., A. Gumy, R. Grainger and U. Wijkström. *The State of World Fisheries and Aquaculture 2008*, Rome: FAO Fisheries and Aquaculture Department. Food and Agriculture Organization of the United Nations, 2008. Print.

Roger & Me. Dir. Moore, Michael. USA: Dog Eat Dog Films and Warner Brothers Pictures, 1989. Film.

Rogoff, Irit. *Terra Infirma: Geography's Visual Culture*, London: Routledge, 2000. Print.

Serpell, James, "People in Disguise: Anthropomorphism and the Human-Pet Relationship." *Thinking with Animals: New Perspectives on Anthropomorphism.* Ed. Lorraine Daston and Gregg Mitman. pp. 121–36. New York: Columbia University Press, 2005. Print.

Shoah. Dir. Claude Lanzmann. Historia. Les Films Aleph. Paris: Ministère de la Culture de la Republique Française, 1985. Film.

The Silent World (Le Monde du silence). Dir. Jacques-Yves Cousteau and Louis Malle. Paris: FSJYC Production, RequinsAssociés, SocietéFilmad, Titanus, 1956. Film.

Slotkin, Richard. *The Fatal Environment: The Myth of the Frontier in the Age of Industrialization 1800–1890.* Norman: Oklahoma Press, 1994. Print.

Sobchack, Thomas and Vivian Sobchack. "The Documentary Film." *An Introduction to Film.* By Thomas Sobchack and Vivian Sobchack, pp. 346–73. Boston: Little, Brown & Co, 1980. Print.

South Park. Created by Trey Parker and Matt Stone. USA: Comedy Central TV, 1997-present. TV Program.

Starosielski, Nicole. "Media under Water: Friction, Flow, and the Material Geographies of Undersea Cables." Doctoral dissertation, Santa Barbara: Department of Film and Media Studies University of California, Santa Barbara, 2010. Print.

Swartz, Wilf, Enric Sala, Sean Tracey, Reg Watson and Daniel Pauly. "The Spatial Expansion and Ecological Footprint of Fisheries (1950 to Present)," *PLoS ONE* 5:12 (2010): 1–6. www.plosone.org/article/info%3Adoi%2F10.1371%2Fjournal.pone.0015143. Accessed 9 December 2010. Web.

Thompson, K.S. "The Fall and Rise of the English Bulldog," *American Scientist*, May–June (1996): 220–23. Print.

United Nations Environment Programme. "Overfishing, a Major Threat to the Global Marine Ecology." *Environment Alert Bulletin*, 2004. www.grid.unep.ch/index.php?option=com_content&view=article&id=73&Itemid=400&lang=en&project_id=7E29170. Accessed November 2013. Web.

The Whale Warrior (aka *Pirate for the Sea*). Dir. Ron Colby. USA: Artists Confederacy, 2010. Film.

10 Cooling the Geopolitical to Warm the Ecological

How Human-Induced Warming Phenomena Transformed Modern Horror

Christopher Justice

The Arctic, Antarctica, and Hollywood share a symbiotic history. Through-out the 20th century, those regions have provided invaluable content for film-makers, but their cinematic depictions have produced a polar mythol-ogy portraying them as impenetrable, exotic, and mysterious. In cinema's nascent years, the brutal inhospitalities of these landscapes and eccentric, rugged characters inhabiting them offered film-makers ideal settings and provocative themes. Originally, documentaries and their journalistic verite were the ideal genre to feature such locales.

In 1919, Frank Hurley, Ernest Shackleton's photographer, organized footage to produce the documentary *South*, and Robert J. Flaherty three years later directed cinema's first full-length documentary, *Nanook of the North*, which featured northern Canada's Inuits and their Arctic ordeals. These documentaries triggered a chain reaction that reverberates today: Polar documentaries continue to reveal the tribulations of heroic adven-turers, but an impressive catalogue of documentaries featuring the natural beauty of the Poles' biodiversity and geography has also been produced, with *March of the Penguins* (2005) and *Encounters at the End of the World* (2007) recently topping the list. However, while the Arctic's natives and geographical appeal first captured film-makers' attention, currently its eco-logical dynamics have become a prescient subject.

One overlooked chapter in polar films' history is the role polar ecology has played in horror. In the 1950s, during the pinnacle of horror's most misunderstood subgenre—the creature feature film, or as Patrick Lucanio describes them, "alien invasion films"—polar landscapes were an impor-tant narrative trope for depicting exotic settings that harboured dinosaurs, aliens, mutant insects, and reptilian beasts (1). In alien invasion films, polar ecology was subordinate to more critical geopolitical issues that dominated many films' narrative locus including expositions of military power; threats from The Other; and collaboration among civilians, scientists, and military personnel. The decade's zeitgeist was not ready for eco-trauma films.

Nevertheless, while 1950s alien invasion film-makers overlooked polar ecological dynamics, they revealed each region's climate and topography, ideal narrative conventions for emphasizing explicit geopolitical themes: American imperialism, Cold War anxiety, rampant xenophobia, and masochistic jingoism. Cold, searing winds; expansive, mountainous terrains; and massive, suffocating snowstorms were visual metaphors of the bitter propaganda wars, unending diplomatic efforts, and hyperbolic fear-mongering that defined America's political stage in the 1950s. The cold, claustrophobic mood that infiltrated post–World War II America helped audiences relate with angst to these frigid, foreign landscapes and their geopolitical connotations.

Unfortunately, cinema's portrayal of the Poles has embedded in our collective consciousness a myopic view of these regions that in recent decades—due to our increasing ecological concerns—is starting to change. These portrayals have implied the Poles' primary commodities are adventure and danger: ideal staples for cinematic commerce. However, whereas polar landscapes in cinematic fantasies produce action and drama, in reality they produce natural resources including water, oil, copper, natural gas, and fish. This illusion has sacrificed an important reality about the Poles' biodiversity, another reason for the deluge of recent eco-trauma films set there to depict these regions' ecology.

Hollywood has depicted the Poles as primitive because they are absent of humans. Their value is anthropomorphic because once humans occupy them their value increases. Sadly, this stance ignores their ecological richness and biodiversity. Furthermore, polar films suggest that only adventurous, grandiose characters can survive their extreme conditions. While not the ideal location for everyone, they are ideal for generals, soldiers, scientists, hunters, anglers, and other stereotypically masculine professions. However, more than four million people live in the Arctic, a diverse geographic region that includes many nations and women. The Arctic is not the No-Man's-Land Hollywood has suggested. Many films have depicted the Poles as ecologically impotent realms, where life is frozen and freakish, a place where only bizarre, extraordinary happenings occur. But since the 1970s, when an environmental consciousness spawned in cinema a proliferation of eco-trauma films—that perspective has evolved.

A perfect storm of ecological traumas has inspired an avalanche of polar films featuring more dynamic ecological themes and crusaders. Gone are polar landscapes' tertiary status, replaced by narratives focusing more on their unique climate, geography, and ecology. Also gone are the stock characters, replaced by politically sophisticated individuals motivated by economic, ecological, political, and spiritual concerns. As global discussions about ecological trauma have infiltrated media, a new awareness of polar landscapes has emerged, resulting in more films across a range of genres featuring polar landscapes. Disney has made numerous films about the region's canine heroes, and other animation studios have feasted on the Poles' potential for children's comedies. Television has joined the bandwagon with *Deadliest Catch* and *Arctic Roughnecks*. This profusion of polar films across many genres and mediums in a short period of time is prophetic.

Because horror films have become a litmus test for global culture's debate on human-induced warming phenomenon, polar films have proliferated in horror. Just as geopolitical tensions in the 1950s appeared in alien invasion films, ecological tensions are appearing in contemporary polar horror films. Starting with films such as *Orca, the Killer Whale* (1977) and Carpenter's *The Thing* (1982), the genre since 2003 has produced numerous polar eco-trauma films. The pervasiveness of polar ecology as a predominant *topos* in horror has transformed the genre.

Polar ecology has established a new visual canvas for horror, ushering in a new era for horror's *mise-en-scène* where sunlit, white, and open spaces are replacing the dark, claustrophobic, manmade places in classic horror films. Natural settings such as tundra fields and icy seascapes have replaced urban and suburban landscapes such as apartment complexes and shopping malls, and the ecological has replaced the psychological. As humanity's understanding of ecology deepens, so does horror's treatment of this science, and because ecologists have deftly explained how the fragility of many landscapes' biological systems have been traumatized, horror has become the ideal genre to exploit those tensions. The pervasiveness of polar horror films is now an index to the Western's world ecological anxiety.

Lucanio argues the "alien invasion film," which includes "alien" species such as giant insects, is a distinct product of the 1950s (1). He concludes that alien invasion films are a "manifestation of libidinal repression" that exposes repressed fears about death, sexuality, identity loss, torture, and other traumatic experiences (5–6). Their success depends on repressed fears: the more repressed our psychological desires, the more resonance they gain.

Because these films combine characteristics of horror and science fiction and exploit repressed fears, Lucanio's explanation of these genre's differences is critical. He writes, "to differentiate horror from science fiction is not to distinguish the two as separate genres; rather, it is to recognize . . . [them as] separate but equal projections of fear" (5–6). Although the genres differ in *how* they project fear, their purpose is to exploit audience repression. Horror films exist in an "alternate world" full of exotic, supernatural realms and strange, unreal characters. In contrast, science fiction films exist in "the continuous world," where monsters originate from real places and infiltrate real settings. Alien invasion films set in polar landscapes are paradoxical because they combine the realism and surrealism of both genres to invoke alienating, anxious experiences. They demonstrate realism because they depict natural settings—the Arctic and Antarctica—that have gained significant media exposure. Additionally, in 1950s alien invasion films, stock footage showcasing U.S. cities and military bases created a sense of cinema verite that infused these films with authenticity. In contrast, they're surreal because audiences' visual experiences of these landscapes were new and sometimes inaccurate. In the 1950s, polar landscapes were foreign and exotic, and stock footage featured the terrain of landscapes or military bases in the lower 48 states. Additionally, the dream-like depictions of monstrous creatures juxtaposed against realistic land- and cityscapes through matte shots and related special effects infused these films with a sense of nightmare and a surreal *mise-en-scène* full of dreaminess and illusion (Lucanio 6, 68).

John Carpenter's differentiations between "left-wing" and "right-wing" horror are also useful. "Left-wing horror" narratives reveal monsters that surface from within communities, facilitating a societal purging and progressive, corrective remediation. "Right-wing horror" narratives reveal monsters that emerge externally, fostering a solidification of traditions and conservative "hardening" of societal values (Jones 146–147). Jones extends Carpenter's ideas: "Narratives in which a threat is used to define or shore up a dominant or indigenous culture . . . might be called 'secure,' while those narratives in which cultural norms are relativized or problematized are paranoid" (146–147). Alien invasion films, which feature external threats from subhuman beasts to showcase the power of existing military resources, are conservative films that foster security.

In the 1950s, the public's repressed fears often emanated from Cold War angst. Writing about horror's "reactionary wing" and "its potential for the subversion of bourgeois patriarchal norms," Wood states, "The political level of '50s science-fiction films—the myth of Communism as total dehumanization—accounts for the prevalence of this kind of monster in that period" (Wood 133–135). Subhuman monsters—with their exaggerated, absurdly inhuman characteristics—reflected Communism's "dehumanizing" features, which made them easy to destroy. Because humans' inability to communicate with these monsters negated negotiation, the only remediation left was annihilation; as Polan writes, any other choice would reflect "misguided weakness" (143). These creatures' reptilian or insect-like features dehumanized them further, tossing them rungs down the evolutionary ladder. They manifested the public's repressed anxieties toward an elusive, invisible enemy: the U.S.S.R. And in the context of Cold War politics, thermal weaponry—a potent metaphor for human-induced warming phenomena—was designed to eradicate these beasts.

In *The Thing from Another World* (1951), thermal weaponry threatens the military, but eliminates The Thing. When soldiers melt ice to unleash a flying saucer, a concerned reporter says, we'll be "turning a new civilization into a 4th of July piece." The reporter reveals how "heat" conjures nationalistic feelings, which represent the military's imperialistic intentions: Military brass want to destroy another civilization—the alien's. Initially, the reporter's concerns are justified: Heat frees The Thing and instigates terror. However, the film advocates using heat to destroy The Thing: The military supports the use of "thermite" bombs; soldiers ignite The Thing with flame-throwers; the lone female character urges the men to "fry" it; and later, they electrocute it. The Thing provides an opportunity to exploit thermal weaponry's power. By film's end, even the recalcitrant reporter is convinced that using thermal weapons was wise.

Warmth in these films means protection; in *The Thing*, the warmth inside soldiers' quarters produces safety and entertainment. Civilian scientists construct a greenhouse—a metaphor for human-induced, utilitarian heat—to further mankind's knowledge: The Arctic's inhospitable geography and climate are no match for their science. Later, The Thing prohibits oil from entering the soldiers' quarters, creating a metaphorical "oil embargo," a popular strategy for freezing "polarizing" states. The Arctic setting exacerbates humans' reliance on heat and reveals how warmth produces civilization. Without human-induced warmth, uncivilized terror reigns, but with human-induced warmth, civilization prospers.

In *The Beast from 20,000 Fathoms* (1953), atomic testing melts polar ice and releases a prehistoric dinosaur. When it appears in New York City, the military uses a radioactive isotope to destroy it. Thermal weapons spawn and destroy the beast; while these weapons produce concern, their redemptive power prevails. In *The Deadly Mantis* (1957), although a volcanic eruption releases the mantis from its frozen slumber, the volcano is still a heat-inducing *natural* phenomenon. The eruption is another reminder of the dangers "warming phenomena" cause. After wreaking havoc on Washington, D.C., synthetic toxic explosives destroy the mantis. And in *The Giant Claw* (1957), an atomic projection device doesn't kill the hideous prehistoric bird, thus revealing atomic energy's limits, but does penetrate the bird's anti-matter force field to allow conventional weaponry to destroy it. These geopolitical implications are powerful: Warming phenomena, whether human- or naturally induced, produce terror—an acknowledgement of its danger—but also alleviate that terror and defeat Others.

However, in *The Land Unknown* (1957) this pattern changes. The film is set in Antarctica, and although the monstrous dinosaurs thrive on heat, that heat is not human induced, but the result of a geological aberration: a polar oasis of "warm water surrounded by a desert of ice." Characters ponder the geological and climatologic implications this oasis holds. Stock footage of polar landscapes includes seals and penguins, revealing the region's biodiversity. Approaching the oasis, the expedition team's helicopter hovers below sea level, but the temperature increases as the altitude lowers, prompting one character to state, "Climate change—one of the main causes of evolution—doesn't exist here." The characters acknowledge that if Mount Erebus erupts, more snow will melt and expand the oasis's boundaries. One character laments, "It's hard to believe that millions of years ago this region was subtropical," suggesting she could wear a bathing suit. Another is struck by the ominous portents this oasis harbours: He asks if it's the "beginning of a new heat wave or the tail end of a million year old cooling off process." *The Land Unknown* foreshadows contemporary horror films' obsession with warming phenomena's impacts on ecology.

Nevertheless, geopolitical purposes inspire the polar expedition. The narrative features a military excursion to "facilitate our Navy's geographical and meteorological mapping operation" and record fossil remains of birds and deposits of copper, iron, nickel, and uranium. Footage of a ship breaking through oceanic ice fields reveals a major geopolitical theme: no terrain can stop the U.S. military's technological, organizational, and scientific power. When the lone female states, "It's overwhelming this tremendous force chewing up the ice," and when her beau replies, "diesel oil," the geopolitical implications surface: The United States will smash anything to pursue scientific knowledge and natural resources. Because polar landscapes were mainly depicted in pre-1950s cinema as impenetrable and inhospitable, they were ideal landscapes for showcasing military might and alleviating Cold War geopolitical trauma.

In *The Beast*, "Operation Experiment" includes a secret atomic testing protocol. The military has overcome the Poles' geography until an explosion producing a huge mushroom cloud causes glaciers to melt (in photographs resembling contemporary footage of crumbling glaciers implicitly educating the public about global warming). This "test" melts ice and unleashes a prehistoric beast. Here, a human-induced warming phenomenon is responsible for a geopolitical threat. This trauma is foreshadowed when one soldier says, "When energy of that magnitude is released, it's never over," and later adds, the "cumulative effects" will continue. The military wonders if this "accident" is the first chapter of a new Genesis or last chapter of an old Genesis, revealing this trauma's apocalyptic connotations. The military's obsession with thermal weaponry is a visual metaphor equating climatologic trauma to a geopolitical threat: A soldier disappears into a snowdrift, the beast's movement triggers an avalanche, and it follows Arctic currents south into New York City. Geopolitics, ecology, and climate change are inseparable.

The Deadly Mantis, the genre's most jingoistic alien invasion film, begins with an establishing shot of a large map, then zooms onto the Weddell Sea, then a tiny island where a volcano erupts, and a voice-over states prophetically, "For every action there is an equal and opposite reaction." The setting shifts to the North Pole, suggesting global trauma, where footage of melting ice, turbulent seas, and crumbling glaciers reveal a warming phenomenon that releases the mantis. The film stimulates our ecological fatalism: Nature's behaviour, and ours too, in one location will impact distant lands.

The film also elaborates on the military's Arctic conquests. A chauvinistic exposition reveals the military's complex radar systems, detailing their three layers' impenetrability. This radar is mankind's ubiquitous protector, surveying deserts, mountains, polar regions, and oceans. Building this elaborate radar system, named Red Eagle 1 ("the sentinel of the Arctic"), under secrecy in dire Arctic conditions was a "desperate gamble against weather" and akin to World War II's Normandy invasion, an event with profound national and global impacts. The film's triumphant music amplifies the project's completion because the Arctic is another U.S. military conquest. Nationalism also soars in *The Giant Claw*. Set in Alaska, a voiceover states, "At the top of the world men struggle against the elements to create a defense to protect freedom." Alaska's polar landscape is ideal for developing extensive radar systems, guided missiles, and bombing planes that showcase the U.S. military's technological prowess.

Because "aliens" target U.S. cities harbouring significant economic, political, or cultural resources, their reigns of horror are terrorist acts that reminded 1950s Americans of their geopolitical vulnerability and the political, scientific, and military resources available to fortify those weaknesses. These films are ideal examples of Carpenter's "right-wing" and Jones's "secure" horror: A threat from the outside is used to assert, justify, and ultimately celebrate existing values and resources. The rhedosaurus in *The Beast* destroys fishing boats and a lighthouse, important components of the U.S. commercial fishing industry, and later annihilates Manhattan's infrastructure, killing 180 residents and causing $300 million in damage. The beast's blood releases a germ that becomes a biological weapon. Later, it crushes a roller coaster in Coney Island's amusement park. Human-induced warming phenomenon can weaken the entire spectrum of American society including food sources (the fishing industry), an economic epicentre (Manhattan), medical facilities, and the entertainment industry (the roller coaster).

In *The Deadly Mantis*, footage of the mantis crawling up the Washington Monument assumes symbolic power: George Washington successfully organized the country's first professional army. However, the mantis mocks America's military and architectural achievements by surmounting that monument so easily. Considering the mantis destroyed the Arctic military outpost earlier, its wrath is aimed at the U.S. military. It leaves Washington, D.C., and travels to New York City, where it finds refuge in the Holland Tunnel, a major transportation portal. In *The Giant Claw*, the prehistoric bird attacks the Empire State Building, a potent symbol of America's economic and architectural prowess, and the United Nations building, the ultimate icon of geopolitical power. Warming phenomena ultimately threatens national identities.

Furthermore, these films' insist science is an omniscient social force that, when collaborating with the military, can combat national threats. Although in many films the military prevails during squabbles among scientists, civilians, and politicians, that science is crucial in geopolitical debates is indisputable. Although these tensions are exploited in *The Thing*, military personnel, scientists, and scholars reluctantly find ways to collaborate and resolve threats. In *The Beast*, although one professor is diagnosed with "traumatic hallucinations" and dismissed as crazy, he recovers and cooperates with military officials. Although chaos is unleashed, the best method for restoring order is to research and develop new knowledge. As Polan argues, scientists in these films are competent: They offer sound solutions, and when they create trauma, it's because they were mad or mistaken (148). Aliens are catalysts that instigate better science. The military facilitates scientific production and inspires it with nationalistic purposes, stark visual and narrative reminders of how the post-industrial-military complex functions. Destroying the monster is not as important as destroying it with U.S. military resources and strength.

Many films conclude with a pervasive sense of harmony and humility: The monster is defeated, factions find common ground, and peace is restored. However, another reason why harmony is restored is because domestic anxieties are resolved. In what Wood calls "revenge of nature" films, monstrous attacks are triggered by familial tensions, including sexual dissatisfaction between spouses, the predominance of patriarchal systems, or the inability to effectively communicate (123). If these factions are metaphorical "families," alien invasion films serve an additional geopolitical purpose: They resolve conflicts within one of America's most potent Cold War weapons, the nuclear family and its domestic front. In many films, each "family's" inability to collaborate exacerbates the monster's rampage. Nevertheless, this fumbling leads to effective communication, understanding, and collaboration between factions, and as Polan notes, 1950s horror films also presented a problem—invading aliens—that facilitated romance and love (149–150).

Sexual tensions are often resolved in these films, female characters duel with "armies" of men, and she—occupying a stereotypically male profession: reporter or scientist—predictably falls for the Alpha male. Banter is laced with sexually charged connotations. For example, in *The Thing*, the lone actress suggests, "If I start burning up again, who'll put out the fire." The two—sexuality and thermal weaponry—are intimately linked.

Furthermore, these monsters' predatory instincts represent uncontrolled male sexuality and patriarchal bravado, so the narratives recall society's need to control patriarchal impulses. Beasts are unleashed because of the military's hubris, which arrogantly assumes Mother Earth can be tamed. Polar regions are metaphorical stages for men to revel in their masculinity, but these regions remind society that masculinity and aggression must be regulated. Masculinity and military power must be balanced with empathy and collaboration.

Polar alien invasion films offer visual representations of the trauma we've created. As Polan notes, the horror is now "part of us, caused by us"; we're responsible for the monstrous. Modern horror films "reject or problematize this simple moral binary opposition" that suggests horror offers only marginal, different manifestations of The Other. Our inability to communicate with aliens is "an inherent part of the human realm itself, not something that assails humanity from an elsewhere." The oppression these aliens instil reminds us "that the very act of constituting another is ultimately a refusal to recognize something about the self" (143–144). Lucanio notes that even when the Earth unleashes terror, we're forced to deal with our mutative tendencies because the Earth's mutative, chameleon-like duality—both creator and destroyer of life—is our duality, a daunting, traumatizing paradox (Luciano 74). Society's repression of geopolitical angst, sexual tension, or identity ambiguity does not deny external forces, but represses internal forces. These films help us revel in oppressive acts of denying ourselves, which is why polar landscapes are central. They're not foreign landscapes, but rather, part of our national landscape *and* our enemy's—the U.S.S.R's—occupying distant, remote parts of our consciousness. Like its Cold War adversaries, America too is destroying living creatures with atomic, heat-generating weaponry (and in contemporary contexts, heat-generating life-styles). These hideous aliens share our history and planet, and through their anthropomorphic tendencies—seeking revenge, desiring survival, and so on—are part of us. Lucanio argues, "The monsters and aliens . . . are best seen not as projections of the repressed unconscious, but rather as projections of the collective unconscious" (20). The trauma these films expose is part of a universal trauma related to death and survival best expressed in the 1950s through geopolitical contexts. Although these films helped eliminate grotesque, fabricated enemies, they paradoxically reinforced our own anxieties and traumas, including the need to destroy.

Throughout these films nascent ecological sensitivities suggest an environmental occurrence in one location will impact another continents away. That complex systems are interdependent is an emerging, yet subordinate theme. So is the notion that cause-effect relationships between seemingly disconnected systems are fundamental to scientifically understanding ecological problems. Every action inherently includes an equal but opposite reaction, and because these films' narratives occur in natural settings, the dawning of cinematic ecological trauma flourished in unprecedented ways with these films. Two pivotal films carried their baton.

ORCA, THE KILLER WHALE (1977)

Writing about mythical landscapes in *Jaws*, Lemkin argues, "America is at best a series of widely scattered and discrepant regions, each with its own unique characteristics, often dependent on the natural topography of the landscape" (322). As land masses our prehistoric forebears traversed, polar regions and the seas that surround them represent our primordial and ancestral past, and few landscapes embody America more than the sea, which served as many immigrants' "first wilderness," the terrain crossed while traveling to the New World (330). The sea throughout history has symbolized the unknown because it's an antithesis to science; the wild sea is "beyond the rule of man, whose influence stops at the shoreline" (323). Consequently, polar landscapes—which combine ice, snow, and oceans—recall universal, prehistoric, and traumatic narrative, migratory experiences. *Orca, the Killer Whale* exposes these traumas astutely.

Andriano argues, "*Orca* is ponderous and discursive like *Moby-Dick*, not exciting and suspenseful like *Jaws*" (29) and "illustrates a cultural turning point: in the 1970s, 'there began to develop a different awareness of whales'—a sympathetic, ecologically conscious one" (16). *Orca* appeared when environmental activism spread into cinema, transforming horror into an ecologically conscious genre and when "Save the Whales" campaigns and their ecological arguments were gaining momentum. Cinema's aesthetic mythos infused eco-horror films with rhetorical power, and science's emerging understanding of ecological dynamics coalesced with that mythos to form a new articulation of the "alien invasion" film. "The camera captures the sublime spectacle of the breaching whale sporting in the waters as impressively as Melville's prose," Adriano writes (33), and ecological themes, not geopolitical, inspired that spectacle. In the 1970s, a new monster had replaced the Soviets: the ecologically ignorant, usually manifested as a corporation or ecologically foolish citizen.

No longer do we have a diverse team of civilians, military personnel, or experienced scientists fighting aliens, but pods of individuals defined by self-interest fighting animals made monstrous by human ignorance. No politically significant cities or landmarks are targeted; instead, battles waged between humans and "monsters" unfold in rural, desolate settings such as oceans, forests, or deserts. Humans still initiate chaos, but terror is unrelated to thermal weaponry; instead, human's economic dependence on and relationship with Nature instigates trauma. Instead of aliens escaping the Poles' harsh terrains, we now have a "monster" luring humans back to the Poles in a surreal migratory experience into humanity's primitive self.

Orca is different than earlier 1970s eco-horror films because its primary focus is on killer whales' ecological and ethological importance. We learn more about killer whales—their communication, predatory, and parenting skills—than we do about animals featured in related films. In contrast to 1950s' subhuman monsters, this killer whale is not absurdly deformed; instead, it's portrayed naturally as a complex mammal with characteristics equal to humans'. However, although scientist Rachel Bedford represents the apex of humans' scientific knowledge of killer whales, her knowledge is limited. Her crusade to educate Captain Nolan about the killer whale's behaviours fails because science is depicted differently.

Unlike 1950s films—where monstrous attacks were surprising—scientists in modern eco-horror films understand ecology and animal behaviour. Or so they think. Unlike in *Jaws*, where the film's soundtrack shrouds the shark with dread and suspense, the music in *Orca* adds a mysterious dimension to the whale's character. The eerie and eloquent orchestral music adds mystery to the whales, which clashes with Bedford's scientific acumen; the whale's aggressive behaviour shocks her because science cannot answer everything about these mammals. However, the problem isn't a lack of science, but rather, the inability to effectively communicate scientific information and persuade others of its relevance. In modern eco-horror films, terror is the result of people—typically with vested political or economic interests—not listening to scientists.

The whale's offensive against Captain Nolan—who killed its spouse and infant while procuring them for an aquarium (Figure 10.1)—is fundamentally ecological. The whale terrorizes a fishing village, and local commercial fishermen have stopped catching fish because the whale scares their quarry.

The whale enhances its attack by destroying the village, and its wrath not only satiates its own lust for revenge, but protects the marine ecological system that supports killer whales. This whale is a metonym for the entire oceanic ecosystem. Furthermore, its unique ability to use landscapes—icebergs and ice—as weapons demonstrates how, unlike in 1950s films, landscapes in modern eco-horror films are crucial.

As the whale lures Nolan north, the film's visual canvas contradicts the dark, claustrophobic *mise-en-scène* of classic horror films. Day-for-day shots featuring sun and glacial whiteness are juxtaposed against night-for-night scenes, giving this antagonistic landscape a schizophrenic "personality." The landscape is as formidable an adversary as the whale. Nolan's friend warns him that ramming into large ice chunks will sink the boat, but Nolan replies, "Ice cuts both ways," implying the whale too must cut through ice for air. However, Nolan's ignorance of polar landscapes and the whale's mammalian features solidify his fate. Ice accumulates on his boat, making walking difficult. Nolan fights the whale, weather, and geography, but is unprepared to conquer them. The whale pushes an iceberg into the boat, triggering a chain reaction that causes an avalanche, killing one of Nolan's crew. The whale's ecological intelligence—not the old fisherman's—wins this battle.

During this struggle, Nolan is monstrous, and the whale is heroic. Bedford's voice-over demonstrates this role reversal when she states, "We were hooked . . . the orca was reeling us in." What triggers this reversal is Nolan's failure to see the whale as a mammal. By calling it a fish, Nolan ignores

Figure 10.1 Captain Nolan (Richard Harris) mercilessly kills a pregnant orca whale, incurring the wrath of its mate, the film's antagonist, which pursues Nolan into Arctic waters.

its similarities with humans: it's not a monstrous other, but rather, a complex being capable of multiple identities: "the ferocious shark, the playful dolphin, and the intelligent human" (Andriano 33). Nolan relates to the whale because he too experienced a spouse's accidental death due to Man's irresponsibility (a drunk driver killed his pregnant wife). However, whereas Nolan never sought vengeance, the whale did without hesitation, earning Nolan's respect, which manifests itself when a close-up of the whale's eye reflects Nolan's silhouette. Nolan's chasing the whale after it destroys his house and injures his metaphorical family (his crew) strikes a powerful chord in the stoic captain, and inspired by the whale, Nolan seeks revenge. During the hunt, the fatalistic Nolan proclaims, "He loved his family more than I loved mine," which is why he allows the whale to lure him into its home territory: the frigid waters of the North Atlantic.

Nolan's transgressions—and Man's—are visually depicted by his "backwards migration" into the North Pole, the converse of Man's migration into "civilization." Cubitt suggests migration itself is an ecological medium. He writes, "Like the great migrations of animals, human migration moves not only biomass but everything that travels with animals: their pests, their dung, their diseases and their genes." Perhaps our values, beliefs, and prejudices migrate with us, and as Cubitt argues, everything projects meaning, including migration. Animals and humans are "senders and receivers, nodes in an increasingly interwoven webwork of communication, whose mediations they perform in their bodies and their technologies" (140).

The film's ecological message is that humans and mammals are equal. Andriano explains, "Anthropomorphosizing the killer whale is not a mindless pathetic fallacy; it is clearly done to emphasize that the orca is a magnificent creature, highly evolved, highly intelligent" (33). Although Adriano suggests we're closer evolutionarily to whales than fish, Shubin argues that we're all descendants of fish (see Shubin 2008). And because fishermen have long held symbolic power in Christianity, *Orca* concludes with overtly Christian implications. Bedford is portrayed in Biblical, Virgin Mary–like attire. After the whale kills Nolan, he descends in a crucifix-like pose into the ocean to remind us of the sacrifices humanity must make to coexist with the animal kingdom. Ecological ignorance has its price.

THE THING (1982)

John Carpenter's *The Thing*, with its ensemble cast, eerie paranoiac mood, and extraordinary special effects, emphasizes popular 1950s creature feature themes. However, these similarities serve more as homage than theme; the film is radically different from the original, and the most noticeable difference is the film's lack of geopolitical messages: no reference to the military surfaces and what the research team is studying in Antarctica is vague. This team is not attached to political or corporate agendas. Unlike the original, American researchers are innocent of liberating the creature and causing terror: American imperial greed does not instigate this film's monster. Conversely, a Norwegian research team's mistake does. While the Norwegians' presence reveals the United States isn't the only country colonizing polar resources, Norway's presence—given its neutrality in geopolitics—reveals the film's focus. Although a language barrier between U.S. and Norwegian researchers results in their inability to communicate interculturally, this point is subordinate. Carpenter is content with exploring psychological, not geopolitical themes; *The Thing* is apolitical but deeply psychological, a potent examination of the 1980s American psyche.

Another difference is the film's Antarctic setting and its effects on humans. *The Thing* occurs in a polar setting, not an American city, and writing the story in a geographically adverse setting intrigued screenwriter Bill Lancaster. Antarctica is more barren than the Arctic, and Antarctica's climate and terrain are portrayed as monstrous. Polar isolation psychologically unsettles the characters before the creature attacks, and several become "stir crazy." They contend as much with the climate and terrain as the creature, and when it infiltrates their compound, the paranoia and claustrophobia they succumb to have already been exacerbated by weather conditions. "Aliens" in *Orca* and *The Thing* prey on humans' inability to cope with adverse climactic conditions.

Dr. Blair is portrayed differently than his 1950 counterpart. While scientists in 1950s films argued with military leadership, they conveyed knowledge and composure amidst chaos. Conversely, Blair suffers a breakdown because he understands the apocalyptic ramifications the creature presents. If it infects more living organisms, the chances of global infestation are statistically horrific. Blair is marginalized and locked in a shed on the compound's periphery. He threatens the team's survival as much as the creature and Antarctica. Blair's plight is a metaphor for science's vulnerability in these ecologically challenged times. There's no happy ending because the creature defeats the research team, and symbolically, science itself: A fatalistic mood overwhelms the compound. Only two men survive, but they understand their futures are bleak.

Furthermore, the creature's shape-shifting—it anatomically resembles the organisms it consumes—is a reminder that the monstrous is horrific because it represents our repressed psyches (Figure 10.2). Because characters attack each other, the creature no longer threatens from the outside, but from inside, instigating a purging of existing values. Carpenter said the shape-shifting motif, part of John W. Campbell's original short story but not the original film, fascinated him. Perhaps that is why he decided not to cast females: to remove sexual dynamics from the narrative. This neutering achieved two purposes: to distance Carpenter's film from the original and to emphasize the psychological implications of a monster and its prey resembling each other.

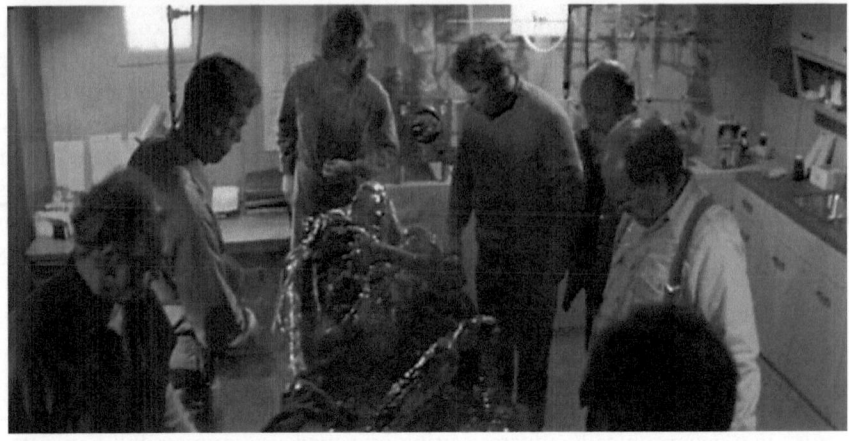

Figure 10.2 The U.S. research team inspects the shape-shifting invader in *The Thing*. The creature is able to infect and take the form of its human hosts.

How does *The Thing* fit into the context of eco-trauma films? Although the film doesn't overtly address ecological themes, *The Thing* does represent three important challenges to ecological ideologies. The creature itself is parasitic: It depends on other living creatures for its survival. This interdependence among organisms is an ecological concept, and Carpenter reminds us how grotesque some organisms that depend on us are. In this compound, the creature needs humans to survive, and in a globally interdependent world, many life forms—from viruses to cruel dictators—need our bad habits and dependence on oil to survive.

Second, the creature is constantly evolving; it defies stagnancy and symbolizes ecology's changing dynamics. A rapidly growing science gaining resonance as a paradigmatic theoretical framework for many disciplines, ecology itself is constantly mutating. One challenge for ecologists worldwide is to harness the complexities of systems theory and interdependent natural systems and to transcend the myopic thinking of binary cause-effect relationships. One variable can produce multiple effects, which can cause a plethora of additional, unforeseen results. The creature symbolizes these mutative characteristics, one reason why the team struggles to understand its properties. Their struggle is ours because the creature embodies core principles of the natural world.

And third, *The Thing* implies that democratic societies with shared values that function as ecological units should be valued over private self-interests. If society is a complex system of interdependent units—individual, family, community, state, nation, planet—and if we expect to survive harmoniously, *The Thing* reminds us that individual instincts for survival will prevail over communal instincts. Americans' penchant for individualism outweighs its need for communal or social cohesiveness; however, that ideology is an illusion. *The Thing* rejects ecological thinking, but by doing so, emphasizes its importance. Because the team implodes from internal anxieties, with each character selfishly pursuing his survival, viewers are reminded that not thinking ecologically in broader, holistic ways can be fatal. Carroll notes that horror films during this period focused on "the recurring theme of survival at a time in American history when economic circumstances have transformed the mere 'bottom line' commendation that 'he/she is a survivor' into the highest badge of achievement" (212). Survival in *The Thing* is equal to "bottom line" profits in business, which is why, as Jones notes, the film depicts the human body and its flesh "as a corporation, a relatively loose agglomeration of discrete units, each capable of acting autonomously, and thus capable of multiple combinations" (Jones 175). This anti-corporate theme is crucial in the context of eco-trauma films because we see in monstrous form how corporations avoid ecological obligations. This "ends justifying the means" ethos undermines shared ecological values. *The Thing* demonstrates that if we do not understand how systems impact each other, we are doomed to a dangerously individualistic, short-sighted fatalism.

ALIEN HUNTER (2003) AND *30 DAYS OF NIGHT* (2007)

Fast-forward to 2003, and a flurry of B-movies—*Ice Crawlers*, *Alien Hunter*, *Deep Shock*, and *Retrograde* (2004)—exemplifies how human-induced warming phenomena have infiltrated contemporary polar horror films. In each, geopolitical themes are subordinate to ecological themes, and ecological and geopolitical challenges clash: Both are part of a global system of interdependent agencies governed by ideological agendas. Ecological trauma has become an integral geopolitical debate, but no satisfactory resolution is imminent. This inability to negotiate shared ecological and political stances emphasizes Cubitt's argument that "emphasizing the connections is not so much an ethical appeal as a mere fact." He adds, "What an ecologically informed concept of mediation has to offer is not a solution but an understanding: that these spheres of activity—economics, politics, education and the rest—are mutually informed and informing" (138). However, in contemporary eco-trauma films, negotiation is impossible.

In *Alien Hunter*, when Dr. Julian Rome melts a mass of Antarctic ice and exposes the existence of an alien creature that carries a lethal virus, Russian and American political and military leaders collaborate—but ignore scientists' pleas—to destroy Rome's polar station and prevent the virus from spreading. The Russians launch a nuclear bomb that destroys the virus and Antarctica. In *Deep Shock*, Russian and American submarines playfully engage in war games while navigating Arctic waters, another reminder of how Cold War tensions have thawed. However, when an unknown source attacks the U.S. nuclear-powered submarine named *Jimmy Carter* (we later discover monstrous electric eels are responsible), the United Nations intervenes to investigate reports of a powerful electromagnetic pulse that increases temperatures in the Polaris trench, threatening to melt Arctic ice and flood various countries. The Research Center, titled Hubris, becomes the centre of an international debate: One G8 leader suggests torpedoes caused the ice to crumble, not melt; another scientist, the rare female, argues the ice is melting due to mysterious energy pulses that warrant further research. She is quickly voted off the G8 committee, and the nations' leaders solve the "warming debate" by exploding the ice shelf with seismic torpedoes, an act of masculine hubris, not scientific veracity.

Polar regions also threaten the world's safety, but aliens are benevolent and forgiving. The scientist Rome is a linguist who studies the communication patterns of alien species, a reversal from the dehumanizing portrayals of 1950s aliens whose destruction was facilitated by our inability to communicate with them. Ironically, characters with similar cultural backgrounds, ideologies, and languages are incapable of effective communication. Although "the dominant mode of thought in the early 21st century is scientific, a diverse and internally conflicted raft of discourses," as Cubitt notes, "science nonetheless invents the math it needs to describe what it finds," and sometimes its discourse communities invent "counterintuitive, even apparently counterfactual constructs" (143). The inability to communicate becomes one of these films' central traumas, and the politicians, scientists, or corporate representatives who facilitate disaster represent the greatest threats. Ideology defines Otherness and prohibits effective mediation. Aliens are not the enemy; we are.

A difference between contemporary polar horror films and their forbears is the presence of monolithic corporate entities and their colonization of polar resources. Corporations have replaced the military by conquering polar landscapes to showcase their technological prowess and business savvy. This occurs in *Ice Crawlers*, where GeoTech uses "experimental drilling technology" that causes an earthquake and unleashes an aggressive prehistoric insect. Throughout the film, anti-corporatism prevails. GeoTech is one "big cost-cutting machine" involved in more conspiracies "than an Oliver Stone movie"; additionally, one character states, "GeoTech decided to slap Mother Nature in the face—start a fight with her and she'll get you—she's big, she's tough, and she always wins." GeoTech is invisible and known only through surrogates; one driller frequently states, "It ain't my problem," echoing GeoTech's corporate irresponsibility, repeated later when a corporate representative reminds graduate students working for him that they signed non-disclosure agreements.

Alien Hunter makes clear there's an oil crisis, and consequently, Axxon Resources (an obvious allusion to Exxon) has established an Antarctic station to genetically engineer foods and conduct research. Axxon's employees are given limited resources, and they are barely debriefed. The food-engineering scientist promotes his agenda through the notoriety and profit it will earn him; the other scientists' inquiries into the weird radio signal—and the alien producing it—are unimportant. And in *Retrograde*, time travellers return to the present to prevent the discovery of bacteria frozen in Arctic ice. A ruthless entrepreneur leads a ship full of specialists, but ignores the captain's pleas about threatening weather and ice flows and urges him to continue, hoping to find evidence of "organics" that will determine if life on Earth "was ceded from another planet." The economic advantage this discovery will provide is obvious. Future profits outweigh present safety or ecological sustainability.

Interestingly, ecology helps characters understand monsters. In *Ice Crawlers*, the first casualty has a blue fluid—a "bioluminescent flora and fauna-eating carrion"—near his wounds. Characters understand symbiotic relationships among animal and plant species, but they don't understand humans' relationships with these species. Their understanding of human ecology is limited. Graduate students criticize the drilling because it's "eroding the ecosystem," and these novices are ironically more ecologically concerned than their mentors, who willingly overlook concerns for profit. However, these students are conflicted because they realize that drilling may perversely benefit science; they weigh which justification for disrupting the ecological balance is more important: to lower gas prices or discover a new prehistoric species. In *Retrograde*, Antarctic biology will determine humanity's destiny. The film argues that once Antarctic bacteria are exposed (by human extraction or, implicitly, by global warming), a global plague will devastate millions. Blindly leading the entourage, a capitalist's greed exposes the bacteria, which kills his crew. As Antarctica is destroyed, polar destruction is aligned with Antarctic biology and capitalist projects to exploit it.

Polar horror films evolve quickly after *An Inconvenient Truth* in 2006 with *30 Days of Night*, *The Last Winter*, and *The Thaw*. Scanning these titles reveals a sea change: Each is uniquely polar and based on meteorological, geographical, climatologic, or ecological phenomena. *30 Days of Night* occurs in Barrow, Alaska, the United States' northernmost municipality. Annually, Barrow is sunless for 30 days, so most residents leave town except civic leaders. When an army of otherworldly vampires uses this climatologic phenomenon to ambush the townsfolk, Barrow's demise is imminent. The film is fundamentally based on ecological, climatologic, or meteorological phenomena. Early, the film's cinematography foreshadows this excessive darkness by capturing numerous panoramic shots of sunlit white horizons along the polar landscape. These shots establish the terrain as a minor character. However, as the barren, overwhelming whiteness connotes innocence, it clashes with the film's prolonged darkness. The cold even reminds characters of death; one says, "That cold ain't the weather, that's death approaching," and this paranoia complements the claustrophobia the cold causes. However, this polar geography also offers unique aesthetic opportunities. Because blankets of snow cover the town, white itself becomes a canvas that accentuates the sprays of blood dappling the town's streets. These sprays ominously function as abstract art, emphasizing geometry amidst chaos. The colour red assumes greater value when juxtaposed against the white, revealing a fascist sense of binary opposites—day/night, good/evil, human/vampire, black/white—where nuance is unacceptable, and reality is paradoxical.

In *30 Days*, humans' relationship with weather allows citizens to survive. While under siege, Barrow's residents realize two advantages: They know the town and the cold. Their relationship with the polar terrain and climate saves them. As one character states, "We live here for a reason . . . because no one can." Another refers to the snowstorm that immobilizes Barrow as a "whiteout," an extreme Arctic climatologic event that slows the vampires and allows the citizens to regroup. Later, one citizen uses a snowplow to massacre several vampires, and their survival strategy hinges on waiting out the darkness and luring the vampires into sunlight so its heat can destroy them. Although killing vampires with sunlight is a popular narrative trope, *30 Days* emphasizes heat's value in a climatologic context.

One cannot ignore *30 Days*'s palpable ecological message. Before darkness sets, one gruff Barrow resident receives a citation for spilling oil, a foreshadowing of the film's conclusion. The citizens' final sanctuary is an oil refinery, a reminder that oil is what attracts people to settle in such harsh environments. However, oil destroys the citizens and Barrow itself. After the vampires break into a pipeline, oil spreads and the town is engulfed in flames. Oil, heat, and fire destroy the town: The natural resource that supports citizens' livelihoods destroys the community, a reminder that Nature creates and destroys. If the Poles are virgin landscapes, we have contaminated them with hubris and greed.

THE LAST WINTER (2006)

The Last Winter changes the focus of polar eco-trauma films with its opening scene. A team of "greenies" and representatives of the oil company North are stationed in the Arctic National Wildlife Refuge (ANWR) preparing for oil drilling. One worker views a North-produced public relations video that celebrates Alaska's exotic qualities and justifies the company's presence in the ANWR. The video demonstrates how propagandistic photographic and cinematic images of the Arctic have produced artificial, myopic perceptions of polar regions. Narrated through North's point of view, Alaska is the "vast wilderness of the north" and a "land of great natural beauty and diversity," suggesting humans should disrupt its biodiversity; Alaska's size and bounty are more than plentiful. Alaska is "rugged country," reinforcing the stereotype that only roughnecks can survive there. Alaska possesses "black gold," a reference to oil's corporate appeal, not its value to consumers. The video's music is triumphant, revealing that conquering the Great White North is another sign of progress for civilization. North's video also reveals why the politically charged ANWR has been off limits to prospectors and oilmen: "Only once have prospectors gained access to this barren landscape." According to the video, in 1986, the Kit Cooperation, "a partnership between native and business interests," explored the region's potential for oil, but that exploration's results have been "carefully guarded ever since." The U.S. Congress protected those secrets and prohibited companies from drilling. However, after a "historic vote" in Congress, that will change. North will transform the ANWR into an oil Mecca that will help America gain energy independence. The video's limited point of view ignores numerous ecological and environmental studies. The video reveals how conflicts between political constituencies such as environmentalists and corporations lobbying for political capital have replaced geopolitical conflicts between nations, and more importantly, how geopolitical issues are no longer embedded in political ideologies such as Communism or Socialism, but rather, in the Earth itself, making the "politics" in "geopolitics" subordinate to the geological.

Arctic landscapes and weather serve a crucial role in this ecological narrative. After the opening credits, a montage of barren polar landscape shots depicting the tundra, sun, and white landscape is offered. The first two shots include a "hot" or "orange" sun, but the next four don't, suggesting a metaphorical polar darkening and ominous foreshadowing. The fourth depicts a piece of wood, and the fifth a set of caribou antlers. They reveal how wood and bone are part of the Arctic's natural biomass and ecology, and that bone and plant matter eventually become fossilized and ultimately oil. The montage's final image reveals a white sun, reminding us that what determines the sun's colour is how the atmosphere scatters its light's wavelengths. The Earth's atmosphere influences our perception of the sun and landscapes, and when atmospheric disturbances surface later due to melting permafrost, these perceptions are distorted.

Throughout *The Last Winter*, strange, unpredictable weather patterns— from hot to cold, blizzards to torrential rains, stillness to sudden winds— characterize the ANWR's climate. The team's cook, a native, says these weather patterns are "like a friend speaking strangely," and the weather is more unpredictable every year. A subjective camera shot offers the snow's point of view, characterizing flakes as a roving spirit. One character dies of hypothermia after walking naked onto the tundra. Later, Ed Pollack, North's lead representative, slips into the ice, also almost dying from hypothermia.

The endless white is a blank canvas that makes objects appear as if they are emerging from the screen. The panoramic white distorts viewers' perceptions because no contours or angles, spatial definition, or background, middle ground, or foreground exist. Some objects lighter in colour blend with the whiteness, further erasing their form and shape. This cinematic phenomenon forces objects within the film's *mise-en-scène* to become objectified, including the environment itself. Because these objects lack form, shape, geometry, and contours, we objectify them to situate them. No longer characters, tractors, station houses, or snowmobiles, they are objects existing abstractly that require us to define their space and purpose. We use these objects for entertainment just as we use the environment. When another representative from North visits the station, he calls the region "ugly as sin," "pure white nothingness," and "the last place on Earth." Footage of the aurora borealis appears, another reminder of the Arctic's atmospheric complexity. When Abby, another North representative, awakens in a medical clinic, a television weather report is broadcast. A doctor has hung himself, and as she exits, a rainstorm occurs as sirens wail, a reminder that the ANWR's toxic atmospheric disturbance is spreading. Her exit is an apocalyptic reminder that Arctic weather patterns and drilling efforts are intertwined with the Earth's ecological destiny.

Animals serve an important purpose in *The Last Winter* because they are symbolic harbingers of impending doom that only the ecologically sensitive understand. While reading scientific data in his outpost, the lead environmentalist, James Hoffman, notices an aggressive crow penetrate his tent. To Hoffman, the crow's behaviour means the "sour gas" (fumes released from the permafrost) may have affected it. The crow's aggression reminds Hoffman of nature's wrath. However, it's the spirit of caribou, ANWR's most symbolic animal, which haunts the team.

When Hoffman leaves his outpost after a bizarre wind gust, hundreds of caribou footprints surround the tent. The "greenies" frequently notice ghostly visitations from herding caribou-like apparitions. During Hoffman and Pollack's trek to a local fort, Hoffman sees the exaggerated caribou spirits, which initially are considered the mysterious Wendigo, but Pollack cannot. Only the ecologically sensitive can "see" environmental anomalies, and therefore, only they can alter the Earth's future.

Although caribou spirits eliminate both men, the spirits' treatment of each is different. One spirit whisks Hoffman away in a sentimental sweep, with Hoffman recalling, nostalgically, his youth, home, and mother. The romanticism underlining his demise is inescapable: As the one individual who understands the ecological ramifications of what's happening, Hoffman does not encounter a violent fate; his individualism, marked by his passion for ecology, unites with the animal's spirit. However, after Hoffman's disappearance, monstrous caribou spirits attack and devour Pollack. Each man's fate is linked to his understanding of ecological sustainability. And Hoffman's assistant, Maxwell, utters the film's most important ecological statement. He understands the essence of oil and how mankind's incessant hunt for it has doomed society. He states, "We shouldn't be here. We're grave robbers. It's coming out from the ground. Ghosts. What is oil anyway but fossils, plants, and animals from whatever millions of years ago?" For Maxwell, extracting oil from Earth is akin to exhuming animal corpses.

The film's emphasis on caribou marks an important tension in eco-trauma films. Maxwell's concerns suggest animals should be left alone because they are different. They deserve separate spaces, and we deserve ours. There is minimal room for collaboration. This separateness, as Mitman argues, is exacerbated by the entertainment industry and reveals a "profound anxiety." Mitman writes, "Our uneasiness with the exploitation of nature for financial gain reveals how much we wish nature to appear pristine, set apart from the hands of man." By watching animals, we objectify them, Mitman argues, and further distance them from us. He adds, "This voyeurism precludes any meaningful exchange" (206) with animals, which is tragic as the popularity and proliferation of entertainment mediums grow in concert with suburban sprawl and population growth. *The Last Winter* argues that meaningful exchanges between animals and humans are possible, especially when humans value animals' ecological fragility.

Hoffman represents the scientific discipline of environmental science, and his ideological perspective is persuasive, assuming a crusading, evangelical appeal. Hoffman is an "oil geek"; he's witnessed every major oil-related global catastrophe, from Iraq's burning of Kuwaiti oil fields to the Valdez oil spill. Ecological disasters define his character. However, his desire to observe the planet's last "pristine, untouched" wilderness prompts his presence in the ANWR. Because Hoffman and Pollack are adversaries, their conflicting agendas represent the film's most memorable moments, adding powerful contributions to the eco-trauma oeuvre. While Pollack wants to know why trucks cannot travel on ice roads, Hoffman replies ironically that it's not cold enough. He explains that sustained subzero temperatures are needed, but recently "the temperature has been all over the place." Their next joust occurs when Hoffman reminds Pollack that he and his team "don't work for North"; instead, they "work for the American people" and will "make sure that North keeps its end of the bargain." Pollack offers a different assessment: The American people want "energy independence," but Hoffman replies that "inflating their tires and caulking their windows" will provide the same amount of oil ANWR drilling will provide.

Hoffman is convincing because he isn't convincing. His data suggests something is wrong, but his caution precludes him from rash conclusions. His flaw is that he's too reticent. As Pollack's nemesis, his composed, methodical, democratic personality conflicts with Pollack's brash, impetuous, dictatorial demeanour. Hoffman believes something is off because "it's in the numbers, but also I can feel it." His fusion of instinct and reason are admirable. However, he doesn't understand the problem, which is refreshing because unlike scientists in related films, this expert doesn't have a solution. He wonders if the environmental anomaly is a sour gas, an atmospheric abnormality, a contagion, or a virus. Because it's affecting everything—the weather, people, and even crows—he doesn't rule out possibilities. Nevertheless, his scientific and ecological instincts are excellent, and we trust his judgment. When an assistant reads Hoffman's journal entry, his voice-over becomes a genuine ecological soliloquy:

> Why do we despise the world that gave us life? Why so alienated? Why wouldn't the wilderness fight us like any organism would fend off a

virus? The world we grew up in is changed forever. There is no way home. Is there something beyond science that is happening out here? What if the very thing we were here to pull out of the ground were to rise willingly and confront us? What would that look like? This is the last winter. Total collapse. Hope dies.

Hoffman's uncertainty marks a new benchmark in how eco-trauma films depict science. We sense he's in control because he relinquishes control. By allowing the polar environment and its ecological dynamics to naturally unfold, he gains deeper insights into humanity's environmental impacts. To stop an ecological disaster, he must allow it to germinate, hoping later to control its spread, a challenging balancing act indeed.

When he shuns Abby, Hoffman's character rejects the traditional romantic role scientists played in these films. Abby works for the "monster," the corporation North. Abby and Hoffman hide their sexual relationship from Pollack because he too desires her. However, as Hoffman's concerns grow, he clashes with Abby; her loyalties are not to him, but to North. She believes he's too "vocal" and suggests his voodoo science is flawed. Later, she suggests Hoffman has nothing more than "a series of alarmist ideas"; she adds, "I can't run based on a hunch. I need evidence." In the absence of certainty, Abby and North argue, the team should remain silent. However, Hoffman understands certainty is impossible. She reminds Hoffman of his confidentiality agreement with North, but he replies, "You can make a real difference here." She ignores his ecological righteousness, pledging her commitment to North. Soon thereafter, Hoffman realizes her relationship with him was a ploy to distract him from North's violations, a dramatic departure from alien invasion films' romanticized past.

Pollack is Hoffman's double. Although North is working "in concert" with environmentalists, their relationship is a sham: Hoffman's assistants wear North jackets, and his notes, once labelled as "property of James Hoffman," are re-titled "property of North Industries." When Pollack tells Hoffman he needs his approval, Pollack says, "I'm not gonna try to handle this with some diplomatic mumbo-jumbo. I'm just going to tell what I need, and I'm gonna expect you to deliver." Hoffman stands firm: "I'm not going to sign something just because you need me to." Pollack calls Hoffman and his team "green flags," conveying the connotation of "red flags" or warnings. The political battle over priorities is clear: What represents the danger? Environmental awareness or oil drilling? To Pollack, Hoffman is a threat, so he monitors Hoffman and his team. Pollack believes their presence is a farce: "North thinks they can win a PR campaign by hiring 'greenies' to do the impact statements; it's a waste of money." When Hoffman explains that the permafrost is thawing, Pollack asks, "What is this some global warming bullshit?" Furthermore, Pollack's dictatorial style is buoyed by a populist, capitalist strain. He reminds Hoffman "The public is with us." His attitude represents the entrepreneurial spirit of North as represented by their corporate slogan: "Trust, risk, and results." When Maxwell's video reveals a caribou apparition, his video contrasts with the corporate propaganda of North. Pollack, in perhaps his most dictatorial act, burns the video.

In the past decade, polar horror films' geopolitical elements have thawed; tensions are now embedded in rifts between ideological constituencies defined by environmental and corporate interests, not political constituencies defined by military or national interests. The alien other is benevolent, and corporations and their proxies are ecology's enemy. These films expose corporations' colonization efforts, irresponsibility, and propaganda. Ecological anomalies and debates define these films' narratives, and humans' interaction with abnormal climate patterns has become a staple. Animals in these narratives have also assumed prominence, and politicized locations have become realistic tableaus where new ecological crusaders with sophisticated knowledge of ecology are pitted against corporate pirates who routinely ignore ecological impacts. Warnings about oil drilling in pristine environmental locations are pervasive. As science's ability to explain warming phenomena has faltered, a new era of ecological spirituality has emerged, transcending militaristic, political, economic, scientific, and environmental agendas while elevating ecological sustainability to a new plateau that values sacrifice as its primary principle. Romances between man and woman, so popular in 1950s alien films, have been replaced by humanity's "romance" with nature. The new polar horror film has evolved from the secure, right-wing horror so popular in the 1950s into a consistently left-wing, paranoid horror that purges society of its most radical sin: ecological ignorance.

WORKS CITED

Alien Hunter. Dir. Ron Krauss. Millennium Films, 2003.

Andriano, Joseph D. *Immortal Monster: The Mythological Evolution of the Fantastic Beast in Modern Fiction and Film*. Westport: Greenwood Press, 1999. Print.

The Beast from 20,000 Fathoms. Dir. Eugene Lourie. Jack Dietz Productions, 1953. Film.

Carroll, Noel. "*King Kong*: Ape and Essence." *Planks of Reason: Essays on the Horror Film*. Ed. Barry Keith Grant and Christopher Sharrett. Lanham: The Scarecrow Press, Inc., 2004. 212–239. Print.

Cubitt, Sean. *EcoMedia*. Amsterdam: Rodopi, 2005. Print.

The Deadly Mantis. Dir. Nathan Juran. Universal Pictures, 1957. Film.

Deep Shock. Dir. Paul Joshua Rubin. DEJ Productions, 2003. Film.

Encounters at the End of the World. Dir. Werner Herzog. Discover Films and Think Film, 2007. Film.

The Giant Claw. Dir. Fred F. Sears. Columbia Pictures, 1957. Film.

Ice Crawlers (aka *Deep Freeze*). Dir. John Carl Buechler. Regent Productions, 2003. Film.

An Inconvenient Truth. Dir. Davis Guggenheim. Lawrence Bender Productions, 2006. Film.

Jones, Darryl. *Horror: A Thematic History in Fiction and Film*. Oxford: Oxford University Press, 2002. Print.

The Land Unknown. Dir. Virgil W. Vogel. Universal Pictures, 1957. Film.
The Last Winter. Dir. Larry Fassenden. Antidote Films, 2006. Film.
Lemkin, Jonathan. "Archetypal Landscapes and *Jaws.*" *Planks of Reason: Essays on the Horror Film.* Ed. Barry Keith Grant and Christopher Sharrett. Lanham: The Scarecrow Press, Inc., 2004. 321–332. Print.
Lucanio, Patrick. *Them or Us: Archetypal Interpretations of Fifties Alien Invasion Films,* Bloomington: Indiana University Press, 1987. Print.
March of the Penguins. Dir. Luc Jacquet. Wild Bunch and National Geographic Films, 2005. Film.
Mitman, Gregg. *Reel Nature: America's Romance with Wildlife on Film.* Cambridge, MA: Harvard University Press, 1999. Print.
Nanook of the North. Dir. Robert Flaherty. Pathe Exchange, 1922. Film.
Orca, the Killer Whale. Dir. Michael Anderson. The Dino de Laurentis Company, 1977. Film.
Polan, Dana B. "Eros and Syphilization: The Contemporary Horror Film." *Planks of Reason: Essays on the Horror Film.* Ed. Barry Keith Grant and Christopher Sharrett. Lanham: The Scarecrow Press, Inc., 2004. 142–152. Print.
Retrograde. Dir. Christopher Kuikowski. Franchise Pictures, 2004. Film.
Shubin, Neil. *Your Inner Fish: A Journey into the 3.5-Billion-Year History of the Human Body,* New York: Vintage Books, 2008. Print.
South. Dir. Frank Hurley. Australia, 1919. Film.
The Thaw. Dir. Mark A. Lewis. Anagram Pictures, 2009. Film.
The Thing. Dir. John Carpenter. Universal Pictures, 1982. Film.
The Thing from Another World. Dir. Christian Nyby. RKO Pictures, 1951. Film.
30 Days of Night. Dir. David Slade. Dark Horse Entertainment, 2007. Film.
Wood, Robin. "An Introduction to the American Horror Film." *Planks of Reason: Essays on the Horror Film.* Ed. Barry Keith Grant and Christopher Sharrett. Lanham: The Scarecrow Press, Inc., 2004. 107–141. Print.

11 Toxic Media
On the Ecological Impact of Cinema

Sean Cubitt

POST-HUMANISM

The term "trauma" includes the sense of an experience too terrible to put into words. For Freud, the trauma of bereavement is typical:

> Reality-testing has shown that the loved object no longer exists, and it proceeds to demand that all libido shall be withdrawn from its attachments to that object. This demand arouses understandable opposition— it is a matter of general observation that people never willingly abandon a libidinal position, not even, indeed, when a substitute is already beckoning to them. This opposition can be so intense that a turning away from reality takes place and a clinging to the object through the medium of a hallucinatory wishful psychosis.
>
> (Freud 244)

The films addressed in this collection address the traumatic state of relations between nature and humans as such a loss. As, in Freud, melancholia is a way of working through the trauma of loss, as an internal process of the ego which, however, is experienced as a (lacking) relation to the world, these films seek to visualize—and sonify—that non-verbal and non-verbalisable experience. They are, however, produced, distributed and consumed inside the same system which they seek to reveal as traumatic—traumatized and traumatizing—in the further sense of destructive, and to a terminal degree. The action of translating trauma into words is therapeutic. It brings the chaos of experience, especially of pain and devastation, into the realm of what might be explained, if never ultimately understood. Perhaps it simply creates a box to put the untellable reality into. And perhaps this is all that this chapter can undertake. Sympathy and explanation do not in themselves change the nature of the brute reality they address, in this instance the unthinkable possibility of planetary ecological collapse. This is the nature of trauma: to make us feel incapable of controlling and changing the traumatic event.

This theory is, on the other hand, based on the common experience of trauma as a problem in memory, a problem of the past. Past suffering cannot be changed, only understood or revenged. In the case of ecological events, the trauma is present and ongoing, and constitutes a problem not in memory but in foresight. If it is the case that humans are those mammals blessed or cursed with the power of foresight, the animals that both plan for and dread the future, then a future trauma is an anxiety which can produce both the kinds of psychotic scenarios investigated in eco-horror films, or the kind of despondency that descends on those whose fear of the future becomes more morbid than melancholy. If the talking cure can offer some relief from the symptoms of past trauma, perhaps the writing cure may provide some relief for the symptoms of apathy and impossibilism that surround our ecological dread.

The power of horror as a genre is to speak the unspeakable. What is remarkable about eco-horror is that often it voices the agony of what has no voice: animals, if indeed they have no voice, but even more so rock, earth, water and air, the suffering Gaia. The parallel with a certain understanding of the perceiving subject in relation to the world is apparent in Wittgenstein's *Tractatus* of 1921:

> The subject does not belong to the world: rather, it is a limit of the world. Where in the world is a metaphysical subject to be found? You will say that this is exactly the case of the eye and the visual field. But really you do not see the eye. And nothing in the visual field allows you to infer that it is seen by an eye.
>
> (Wittgenstein 57)

The self that witnesses is not in the world it witnesses. This is the basis of realism for Wittgenstein, the subject as "point without extension" and the reality which coordinates with it. What makes consciousness possible—a world, my world—is external to the self. This thinking of the world apart from the self is disturbing: It evokes the world that will continue to exist after my death. Death, for example, in Heidegger is responsible for that aspect of human life which confronts mortality with life, which urgently thinks, writes, acts in the knowledge that there is a finite limit to life. But death, Heidegger holds, is strictly unknowable (the deaths of others are knowable but my own is not); and likewise a world without me to know it is not imaginable. But this is exactly what horror confronts as a genre: It works through the ego-centrism of mourning by confronting the actuality of death, the actuality of a world in which "I" am not.

This problem is also the problem of thinking Gaia as agent in history. Ecosophical thinking in recent times has been given a spur by the critical theses known as speculative realism. One leading figure, Quentin Meillassoux (2008), begins his major work with an argument that postmodern thought is dominated, ironically enough, by a metaphysical theme which he calls correlationism: that the work of thinking, naming, describing the world changes the nature of the world, and that the theory that there is no given world is condemned to see the world as a product of subjectivity (individual, social or world-historical like Hegel's *Geist*), abandoning all hope of understanding the world's independence of and autonomy from the human species. Here again the horror genre experiments with this contradictory terrain: a world that has no part of human meanings. Freud would have understood this as a projection onto the world of the withdrawal of affect characteristic of mourning; we should try to understand it simultaneously as a cry of anguish, a scream of revenge, and the weeping of the bereaved, all expressed by a world which, while taking on the role of subject, takes it on only in regressive form of traumatized subjectivity. What is remarkable, in terms of speculative realism, is the inhumanity of the consciousness projected in the eco-trauma genre.

Because what is at stake in both Wittgenstein and Meillassoux is indeed consciousness. Both propose a world without consciousness, and in both cases the tendency is towards a philosophical understanding no longer dependent on human consciousness. Implicit here is the idea that the attributes of consciousness, including intentionality, are no longer considerable within a rigorous philosophy. This poses a serious problem for cinema and before that for photography, where the processes of recording have always involved both the intentionality of the director, the camera operator and the contingency of the world (on this as a property of early cinema see Doane 2002). Knowledge is dependent, for both Wittgenstein and Meillassoux, on being, not on consciousness. But implicit here, again, is the belief that consciousness is individual. The lesson of eco-criticism is instead that consciousness is not only social but tripartite: a human element, a technological and a physical (in the older sense of the Greek word *physis*, nature as both organic and as the laws of physics). The laws of physics strike humans as contingent: Without the internal necessity of logic and reason, they impinge on human freedom. Similarly the affordances of technologies are not only allies (tools) but limitations we struggle with and against, seeking mastery over our own inventions. In the case of cinema, consciousness—in the form of the intentions of film-makers—constantly struggles with the natural and technological worlds which are integral to the work of making films. We can then neither abandon consciousness, nor give it the highest seat in the order of domains which together form the complex of the cinematic apparatus. This apparatus, eco-criticism forces us to accept, is not simply an assemblage combining human and technological elements, but one in which physics, nature, plays its own rich part. Just as we can exhaust crews and wear out equipment, nature is also engaged in the labour of film, not just as something that sheds its image onto the filmstrip. It is a material, a material transformed in the cinematic apparatus, just as technologies and humans are.

Many of our films speak directly or indirectly to this triad, from *Peeping Tom* (Michael Powell 1960) to *Christine* (John Carpenter 1983) and more recent variants like *The Day the Earth Stood Still* (Scott Derrickson 2008). What I hope to do in what follows is to indicate that it is not only in narratives and spectacles that cinema lives out the problematic relations between human, technological and natural, but in the very machinery through which we produce our films. I want also to emphasize not only that there are severe problems here, but that they can be understood in useful ways, that is to say, ways which may help us find solutions to the problems being raised.

Despite the frequent description of the information society and the knowledge economy as "weightless" and "immaterial," there is a very specific weight and materiality to the environmental footprint of contemporary media. Indeed, it may well be the case that, in a history of toxic materials and profligate use of energy to produce the media artefacts of preceding ages, the transition to digital has made the footprint of our media heavier than ever.

The extraction of materials, the production processes, the use and recycling and the energy signatures of each of these steps in the life-cycle of digital devices implicate media technologies in both the natural environment and human societies. A core lesson of scientific ecology, especially in the age of anthropogenic climate change, is the intertwining of all forms of life. This mesh of interwoven impacts and influences must include the technological. When Deleuze and Guattari refer to the machinic phylum, it is in order to emphasize that our devices, while they may be of a different order to living creatures, are also in their own way a physical order, agents in the same way that people and forests are. The three phyla, human, natural and technological (political, technical, physical) are inextricably interwoven even in their differences. Those differences feed our fantasies of apocalypse: the revenge of the virus (*Doomsday* 2008), the revenge of the green world (*Godzilla* 1998), the revenge of the technological (*Eagle Eye* 2008). We work through fantasy to maintain the borders between the human and the non-human. But just as the future of human society depends on opening borders to the free flow of people (just as they are open to the flows of commodities and finance), so the future of the planetary environment demands that we make increasingly porous the boundaries between the phyla.

Abizadeh (2008) demonstrates that the coercive policing of borders contradicts the principle that "democratic legitimation requires that coercive laws be justified to all those subject to them." Our thinking of democracy excludes both our machines and our environment, just as, for centuries, it excluded women and the landless poor. The exploitation of the poor, the devastation of the environment and the enslavement of technologies are of one order, as I hope to demonstrate in the following pages. They too are subject to coercion by an increasingly unified system in which they have no say. From mining to fabrication, from content production to chip design, from hard media to downloads, from handhelds to urban screens, there are environmental implications which reverberate through the three phyla. The political solutions that we will need cannot be exclusively human: This is the message of the displaced subjectivities of the horror genre. Nor can the meaning of sustainability be confined to sustaining human life, at the expense of the other phyla. Such a politics will have to include physical, technical and human dimensions if they are to be political, which is to say if they are to bring about change through genuinely open consideration of the fundamental question of politics, what is to be done?

THE SPECIFIC GRAVITY OF CINEMA

The American Humane Association's Guidelines for the Safe Use of Animals in Filmed Media, the industry-sanctioned oversight for animal welfare in the US film industry, only achieved that official status (in the Producer-Screen Actors Guild Agreement) in 1980, although it has organized for animal welfare since the 1940s. The current guidelines extend the duty of care to insects, defining "animals" as any sentient being. Many audiences would be appalled to think that once-common practices like trip-wires to force horses to fall were still in use today. The same cannot be said for our awareness of the impact of film-making, distribution and consumption on the physical environment. We expect predator-prey behaviour or animal fights to be simulated, but are less concerned with explosions, fires and other on-screen damage, and even less so about the footprint of crews and equipment in urban, rural or wilderness areas. Perhaps we cannot be expected to offer the same level of care to the non-sentient world; perhaps we cannot expect to care about everything. Yet it seems anomalous to care so little, and to know even less, about the environmental challenges produced by the film and television industries.

A critical difference between traditional analog film-making and the new digital tools is that analog equipment tends not to go out of date at anything like the same pace as digital. A 35mm or a 70mm camera might have its motor reconditioned, and require regular servicing, but with a decent set of lenses and a competent user, it is built for a lifetime. The aspect ratios established over the years since the adoption of Academy Standard in 1932 are mostly served by the use of aperture plates, not a fundamental re-tooling of the entire industry. Cinemascope requires only another aperture plate and an anamorphic lens: The film strip and its travel remain of the same size, with the same number of sprocket holes, travelling at the same speeds through the gate, as they have since the introduction of sound. Analog cameras are not especially adversely affected by heat, cold or dust, like so many digital devices. New film stocks are designed to be used in old cameras without requiring fundamental design changes, only filters and lens coatings. Old cameras had for decades a healthy afterlife in the lower echelons of the industry, in film schools and in the amateur market. Yesterday's digital kit is as popular as yesterday's newspaper, and far less tractable than analog cameras to being broken down for spares.

Principles for Environmentally Responsible Screen Production published by Greening the Screen of Aotearoa New Zealand, and the *Green Practices Handbook* from Canadian Green Screen Toronto, provide fundamental advice on how to minimise the impact of power-use, generators, transport, sets and to some extent consumables, especially paper and film stock. The Canadian publication warmly recommends careful recycling of batteries and the use of hard-drive storage and digital cameras, concerned for the wastage of used film in the industry. Missing from this computation is the cost of construction of batteries and cameras. One of the more widely used high-definition digital cameras is the Red One, powered by a V-mount Lithium-ion (Li-ion) battery. The major component of the battery is, unsurprisingly, lithium, the richest reserves of which occur in the vast Bolivian salt lake Salar de Uyuni. Indigenous Andeans have successfully fought off attempts to extract the mineral by foreign companies. Until such time as these reserves are opened up, global lithium production is centred in brine deposits in the Southern Andes of Argentina and Chile, described here by London *Daily Mail* journalist Dan McDougall (2009):

> In the parched hills of Chile's northern region the damage caused by lithium mining is immediately clear. As you approach one of the country's largest lithium mines the white landscape gives way to what appears to be an endless ploughed field. Huge mountains of discarded bright white salt rise out of the plain. The cracked brown earth of the site crumbles in your hands. There is no sign of animal life anywhere. The scarce water has all been poisoned by chemicals leaked from the mine.
>
> (McDougall)

The *Daily Mail* coverage goes on to record a visit by a Chilean delegation led by Guillen Mo Gonzalez to the Bolivians living near the Salar, warning them of similar environmental degradation, and especially of the loss of water used, evaporated in huge quantities in the mining process. Ironically, the McDougall story suggests that the Bond movie *Quantum of Solace* may have derived its plot about the theft of water from just such ecological disasters. Because Li-ion batteries are also crucial for electric and hybrid cars, the large reserves available are even now being parcelled out among hungry corporations planning for the changes in energy use after peak oil. Covering the Chilean industry, *Forbes* magazine, like the *Mail* not known for left-wing sympathies, notes that "nothing grows in the heart of the Salar de Atacama" (Koerner 2008). Surely, salt lakes do not harbour vegetation and the wildlife that accompanies it. The possible function of solar reflection from lakes like the Uyuni, the size of Northern Ireland, doesn't count; nor does the sheer geological beauty of the place, the culture of the indigenous people who live by its shores, or the potential to disrupt the role the mineral deposits play on the lower slopes of the Andes, or the high likelihood that extraction will take place in third world conditions involving dispensing with vast quantities of the chlorine with which the lithium is bonded. The cleanliness of lithium in Red batteries as much as in electric and hybrid cars will be purchased at the expense of the people and the environment of the Southern Andes.

As to the preference for hard-drive storage, some attention needs to be paid to aluminum, the largest component material by weight used in their construction (Idema 2009). One of the most abundant of minerals, aluminum ore bauxite is, however, extremely strongly bonded with oxygen and requires temperatures in the region of 950 degrees Celsius to be extracted. While many of the world's bauxite mines are providentially near hydro-electric plants, China, with almost a fifth of world production, is not so blessed, and working conditions are often unpleasant. On the positive side, aluminum is an eminently recyclable metal. Other hard-drive components include iron, plastics and small quantities of rare earths (lanthanides) to which we will return. Hard drives are impressively carbon neutral, perhaps especially in comparison with film, which requires both petroleum products for the plastic base and significant amounts of silver, which can only be recovered with difficulty, in photosensitive salts. Problems with hard drives are not then so much about the materials as about energy use. HDDs (high-density disc drives, as used in high-definition electronic cinematography) spin at great speeds, and while acoustic design has improved dramatically in recent years (Seagate 2009), the heat signatures of HDDs have remained extremely high, as have their need for the electrical power to spin them, bringing us back to the battery problem, and indicating the degree to which the individual elements of the cinematic apparatus need to be understood in their connectivity with the whole ensemble.

HDDs are also intrinsic to the post-production sector. Our research on the *Lord of the Rings* trilogy (Cubitt and King 2008) gave storage figures rising from 7.5 terabytes for the first film to 72 for the third. Production company Weta's digital equipment base included more than 250 graphics workstations, and more than 2,000 Linux and Apple computers, with a handful of specialist devices thrown in, all connected to a 10 gigabit backbone specially constructed to facilitate the production by New Zealand Telecom (Figure 11.1). The scale of these production pipelines, unusual at the time because of the large scale and geographical location of the production, are not unheard of in the age of the blockbuster film. Not only does post-production require a large number of computers, the structure of sub-contracting shots and sequences to specialist effects houses, sound studios and grading labs increases the absolute numbers beyond what can be accounted for in the offices and studios of the central production company. In the case of *The Lord of the Rings*, more than 50 companies on three continents had significant roles in the production, in many instances duplicating or adding to the storage and processing requirements of the films, and requiring an impressive investment in communications infrastructure. The distributed nature of film production in the 21st century accelerates production, especially in this instance where the international dateline meant that the day's work in Aotearoa could be processed overnight in Europe and North America, ready for the next day, in a 24/7 schedule. Shuttling this quantity of data on planetary scales also requires internet servers capable of handling gigabit traffic, again a common occurrence with runaway productions. The energy usage and heat emissions of such distributed networks are not calculable in any simple way: Each firm bar the central production company typically works on a number of projects simultaneously. Suffice it to say that what is saved in the waste of silver may well be lost in the carbon footprint of post-production.

Digital cinema takes on new weight in distribution. Traditionally the least examined sector of the industry, distribution in the early 21st century is a multi-format process of delivering product to markets in a huge variety of forms. Recent blockbusters like *Watchmen* (2009) are completed with theatrical IMAX versions as well as the standard polyester road show print. But they are also shipped to airlines and in many instances to DVD and Blu-ray presses, well ahead of their disc release, in order to head off threats of preemptive piracy. Theatrical releasing now constitutes a diminishing proportion of total revenues, with TV sales, rental and sell-through recordings, airline and hotel screenings and downloadables for portable screens making up increasing percentages of revenue. The art of distribution is the combination of speed and delay. Price is determined as a function of time: Distributors charge more for swift delivery, and less for delayed delivery. It does not

Figure 11.1   The WETA Digital Data Centre in Wellington, New Zealand, used its PCS100 AVC supercomputer and 40,000 processors to render digital imagery in *The Lord of the Rings* (2001, 2002, 2003), *King Kong* (2005), *Avatar* (2009) and *The Adventures of Tintin* (2011), among many others.

include externalities like the environmental footprint of the various modes of technology deployed in distributing films.

How might we account for these externalities? Almost all contemporary feature films go through a stage of digital intermediaries, so-called even when the source footage has been shot electronically. It is then printed to a variety of formats: IMAX and 35mm film, various forms of disc. But in increasing numbers, films find their way to audiences without hard-print copying, either through digital delivery to cinemas or via downloads. Since the latter form an increasing proportion, and are likely to become standard, we should consider the externalities of electronic delivery. Even when distributed through closed intranets or proprietary satellite inks, films (and associated flows such as the revenue from rentals) pass through servers, also known as data centres. According to the US Department of Energy (2008), "Data centers used 61 billion kWh of electricity in 2006, representing 1.5% of all U.S. electricity consumption and double the amount consumed in 2000. Based on current trends, energy consumed by data centers will continue to grow by 12% per year." In 2006 Boccaletti and colleagues at consulting firm McKinsey estimated that IT manufacture and use is responsible for 2% of global carbon emissions—the same amount as the airline industry—and is heading for 3% by 2020, when it will be responsible for the same amount of carbon as the United Kingdom produced in 2008. The authors argue that "the fastest-increasing contributor to emissions will be growth in the number and size of data centers, whose carbon footprint will rise more than fivefold between 2002 and 2020" (Boccaletti, Löffler and Oppenheim 2). One of the drivers in this movement is the switch from broadcast media to Internet as the preferred delivery mode for rich media like films and other audio-visual material, including legal transmissions, illegal file-sharing and quasi-legal forms like user-posted video on YouTube. It needs to be added that media corporations are increasingly using video-sharing platforms to entice purchasers of tickets or discs, or using similar technologies on their own sites and content-aggregators like Apple's QuickTime trailers portal. We certainly cannot blame the film industry for using all available media.

But the wider apparatus of film also includes its consumers, vital to a web 2.0 and increasingly wireless generation of online business. In their efforts to stem illegal copying, the industry has been a front-rank player in governance fora where issues of digital rights management are in play. Central the concept of intellectual property they draw upon is the idea that films are property. On one hand, this means that, unlike other things we buy, we are not permitted to sell or even give away films, which under copyright law remain the property of their producers. On the other, it means that individuals are expected to buy a copy of the film for their own use. This two-pronged assault leads to a massive hoarding of legal and—in a mirror image of the legal system—illegal copies. It is this storage, powered by the model of private ownership underpinning Hollywood's vision of copyright, which makes such a significant contribution to the soaring energy demands of the server farm industry.

Optical storage media like Blu-ray and DVD are typically made of lac-
quered aluminum with an ArInSbTe alloy recording surface comprising
silver, indium, antinomy and tellurium. Of these materials, indium is on
the critical list, with one source indicating that global supplies will run out
within years:

> Armin Reller, a materials chemist at the University of Augsburg esti-
> mates that we have, at best, 10 years before we run out of indium. Its
> impending scarcity could already be reflected in its price: in January
> 2003 the metal sold for around $60 per kilogram; by August 2006 the
> price had shot up to over $1000 per kilogram.
>
> (Cohen 2007)

By September 2009 this all-time high price had dropped to $520 per
kilo on the European spot-markets, but had begun another rise (Desai and
Maclellan 2009), perhaps as a result over Chinese government fears of
pollution spreading from smelters (Tan 2009) threatening to slow production in the country, which has 60% of world production (Indium itself is
non-toxic, but some of its compounds are poisonous, and others possibly
carcinogenic). Also used in indium-tin-oxide (ITO) coatings for flat-screen
TVs, computer monitors and increasingly new generation handhelds like
the iPhone, indium is, with gallium, also a crucial component in current and
next-generation solar panels, Environment Victoria (2009) is one of many
agencies pointing towards the necessity of recovering indium form the recycling of abandoned digital machines; according to United States Geological
Survey (2009), now a major source of indium. But as we will see, recycling
poses its own challenges, even when, as in the case of indium, the price of
recovered metals is exceptionally high. The same may become the case with
tellurium, also used in the alloy surfaces in optical discs: Rarer than plati-
num, tellurium is the ninth rarest metal on the planet, and also in demand
as a semi-conductor. Silver is, surprisingly, almost the least valuable metal
used in making DVDs.

Despite the value of these metals, recovering them in the tiny quantities
used in each disc is unviable economically. Discs are cheap—so cheap they
are often used to scare birds in gardens. Worse still, optical discs have very
short life spans as archival media: scarcely more than five years, compared
to the much longer life of film stock, now commonly used in the industry
to store the raw code from which key digital properties have been built.
Like so many of our digital media, discs seem to have the kind of built-in
obsolescence we mock in the automobile industry of the 1950s, but which
we accept and even celebrate in the rapid churn of computers, laptops and
personal digital devices of all kinds. Thus whereas, for example, cathode ray
tube (CRT) televisions and monitors were energy intensive in use, replacing
them with lower energy LCD screens leaves the problem of recycling the
older technology. Plasma screens use almost as much energy as CRTs, and

have similarly high heat signatures, a common property in desktop computers and servers. Accelerating innovation cycles in the electronics industry are major drivers of energy and materials use in manufacturing and the physical distribution of both parts and finished machines, of materials extraction and depletion. Many component materials, such as the arsenic used in doping semiconductors, appear in tiny quantities in finished goods, but have to be refined and amassed in extraction and manufacturing, where seepage into water tables is a major problem, causing species and habitat loss as well as toxic effects in human workers.

Screen technologies, critical to the new economics of distribution, also display the effects of private ownership. Cinemas, in their popular heyday in the 1940s, offered one screen for thousands and even hundreds of thousands of viewers each year. Since the advent of domestic television, and increasingly with the rise of "third screen" handheld devices, we have not only a family screen but a proliferation of personal screens as well. This atomization of audiovisual culture has not only cultural but environmental effects. While it makes sense for industry to proliferate more and more screens, with faster and faster upgrade and replacement rates, the demands such developments make on the environment is chilling.

There are geopolitical consequences too. Take, for instance, lanthanides, the rare earths required for both LCD screens and the batteries for hybrid cars. China, again, dominates global production of these metals, also known, for obvious reasons, as rare earths. China's public strategy for recovery from the global financial crisis includes moves into high-tech and eco-friendly product manufacturing, leading the government of the PRC to look for major shareholding in mining corporations in the rest of the world. Frequently rebuffed by nationalist sentiments in countries like the USA and Australia, China has made major inroads by stalking smaller, less publicly sensitive firms. Thus China secured in 2009 significant shares in Arafura, an Australian mining company, to help bolster its global dominance in the rare earths field (*The Australian*, "China Builds Rare-Earth Metal Monopoly," 9 March 2009). In few cases is the intense imbrication of geopolitics, digital economics and environmental catastrophe more apparent than in the case of tantalum, a major mineral in the manufacture of mobile devices. Known colloquially as coltan (from its common occurrence in conjunction with a similar ore, columbite), tantalum mining in the Eastern Congo is associated with child labour, possibly under conditions of slavery, in a region where there is little wood for heating and cooking, driving workers to clear rainforest on the Rwandan border which is one of the last refuges of the mountain gorilla. Funds obtained for smuggled tantalum are then widely believed to be used to fuel the civil war which has ravaged the country since 1998, and in which millions have lost their lives (Hayes and Burge 2003, UN Security Council 2003, World Rainforest Movement 2003). The World Wildlife Fund (n.d.) "estimated that over 10,000 people moved into the Kahuzi-Biega National Park to work in the mining industry." While there is little hard evidence that gorillas in this world heritage reserve have been killed for "bush meat," habitat loss and disruption of mating, feeding and child rearing have been significant in reducing numbers of the animals.

The active audience theory of reception needs to bear in mind that the media we use to receive are not only cultural but also economic and political, and that they are far from weightless or immaterial for the people and places that they come from, and indeed the people and places where they go to. The final stage of the apocalypse of digital media is in the recycling sump villages of Nigeria, Kenya and Southern China (Basel Action Network 2002, 2005). The wealth of toxic materials in waste electronics, from toner cartridges to old CRTs, making their illegal way to these dumps include Polychlorinated Biphenyls (PCBs), lead, zinc, cadmium, arsenic and many others. Because recycling is such a low priority it uses only the cheapest form of energy: the manual labour of the very poor (Figure 11.2). Basel Action Network suggests at least 100,000 are employed in one rural area of southern Guangzhou province.

The work is done with only basic hand tools, and often involves uncontrolled burning of worthless but toxic plastics to recover the metal parts which can be sold on. Cancer rates are extremely high, as are other toxin-related illnesses. The agriculture of the region has been harmed by pollution

Figure 11.2   E-waste dump site in Ghana. Photo by Basel Action Network. Copyright BAN 2009.

of the water table. The Basel Convention bans the export of toxic waste, but is widely ignored (the USA has yet to ratify the convention). According to the Electronics Take Back Coalition, over 200 million computers and peripherals are disposed of every year in the US alone. Of these a large majority go to landfill, only about 18% being recycled. Of these, however, the majority are exported. With the mass expansion of Chinese and Indian economies, more and more computers are in use, and more and more enter the recycling chain. New machines are increasingly avoiding at least some toxic elements in their construction (Nimpuno, McPherson, and Tanvir Sadique 2009), but old machines are the ones being recycled, and the incessant innovation is what drives them out of use and into the recycling villages.

These scenarios of extraction and manufacture, use and recycling create landscapes to equal the infernal vision of the Plains of Mordor. The irony is that, for the most part, eco-apocalypse movies concentrate on urban scenes of filth, toxicity, collapsing environments and mutation, while in reality these already exist, but in remote wildernesses like the Southern Andes, or in once rural and agricultural villages of the global South. The digitization of cinema is far from alone in contributing to the environmental degradation caused by the digital industries as a whole. But it is not, and cannot afford to pretend to be, innocent in it either.

TOXIC AVENGER

The task of eco-criticism is to find solutions, not to create despair. However, Greens have ceded major ground to either eco-fascism or a kind of defeatist pragmatism. Eco-fascism declares a state of emergency (Bookchin 1987), suspending all other considerations in the interests of meeting the crisis. In this it adopts the tactics of an increasingly presidential style of rule, where the executive has extended its powers at the expense of legislatures (Sassen 2006), while attacking any other agenda as irrelevant, in the manner of the war on terror (Agamben 2005). The pragmatism of defeat similarly adopts the rules of the existing regime, suggesting either shopping or voting as the only routes to change (see, for example, Barry 2008, Heartfield 2008). Both of these are based in the forms of the very regime of political economy that has created the problem: in the terminology advanced by Jodi Dean (2005), communicative capitalism. The term is in part descriptive, in part ironic.

Communicative capitalism saturates the world in chatter, a commodified information exchange in which every item is exchangeable for every other, so that my opinion is worth your opinion, news of the price of indium is equal to a download of the latest track from U2, and all information is exchangeable for cash, a medium without a message. At the same time, the political forms typical of communicative capitalism are anything but communicative: Politicians claim to "hear" opposition, but decreasingly both to answer, and even less to debate with it. Participation is translated into choice, an equation which

> treats commercial choices as the paradigmatic form of choosing. In so doing it displaces attention from the fact that the market is not a system for delivering political outcomes . . . Political claims are partisan claims made in the name of and on behalf of a larger group, indeed, of an all that can never be fixed or limited (and so remains non-all), a group perhaps best understood as composed of anyone. Such claims are general rather than individual, and they require those who make them too thin beyond themselves as specific individuals with preferences and interests and consider what is best for anyone.
>
> (Dean, *Democracy* 22)

The claims that democracy makes to represent anybody are, under contemporary conditions, deeply circumscribed. In particular they discount the relevance of non-human actors. When Greens seek to raise awareness of consumer choices and political consciousness, they enter the same circulation as all the other human voices, individual or social, accepting in effect the voicelessness of the wider environment, and the foundational role of shopping as our sole vehicle for making known what it is that we want.

This is how eco-horror enters the dialogue, by asserting that exclusion of the green world from democratic politics destroys the claims of every democracy to universality. Asserting meanings which, in their non-human origins, appear as horrifying, they assert the broken nature of any claim to universality derived from an exclusion. While the Deep Ecology movement's use of direct action is alluring, it is so in the same way that contemporary Hollywood is deeply tied up with narratives of revenge. Revenge too has the feeling of a politics without dialogue, which is no politics at all. The pleasures of horror may perhaps be understood as a fictional working-through of the scenario of the obsessive melancholic analyzed in the essay from Freud previously cited:

> If the love for the object—a love which cannot be given up though the object itself is given up—takes refuge in narcissistic identification, then the hate comes into operation on this substitutive object, abusing it, debasing it, making it suffer and deriving sadistic satisfaction from its suffering. The self-tormenting in melancholia, which is without doubt enjoyable, signifies, just like the corresponding phenomenon in obsessional neurosis, a satisfaction of trends of sadism and hate.
>
> (Freud 251)

If, on the one hand, horror spectatorship places the viewer in the masochistic position of identification with the victims of horror, on the other it displaces the role of abuser, debaser and sadist to the avenging powers of natural forces trampled into destruction by humanity. Freud's theory points to the regression of libido into the ego itself, instead of being turned out towards other subjects and the world. This is exactly the condition of a population without politics.

In later writings, Wittgenstein argued "If a lion could talk, we would not understand him" (Wittgenstein 1968, 223). The world not only speaks but roars in our ears, in a tempest of storms, collapsing glaciers, forest fires, mudslides . . . and yet we *will* not understand. Wittgenstein's point concerned the incommensurable nature of different modes of language. That this is integral to public life is clear from the example of politicians unwilling to engage in debate, and devoted instead to consensus derived from statistical management of expectations, to communicating policy, to "sharing" their own vision: effectively to solipsism. While linguistic philosophy might hold this as a permanent and universal condition, political philosophy cannot. It must undertake to find ways to bring the human and the lion into dialogue.

In the Conclusion to his *Politics of Nature*, Bruno Latour asks,

> What if freedom consists in finding oneself not free of a greater number of beings but attached to an ever-increasing number of contradictory propositions? What if fraternity resides not in a front of civilization that would send the others back to barbarity but in the obligation to work with all the others to build a single common world? What if equality asks us to take responsibility for nonhumans . . .?
>
> (Latour 227)

These questions, drawn from the French revolutionary slogan of *Liberté, fraternité, égalité* open the possibility of another mode of democracy, one that genuinely attempts universality beyond the restricted universals of shopping and voting. In film studies, however, we have become nervous of one of Latour's terms, "taking responsibility," for which we will always read "representing." As long as humans represent, speak on behalf of, we will follow a politics of sustainability whose goal is only the long-term sustenance of humans. Only by taking a step back, by granting autonomy to the rest of the planet, will we come to a politics—and an economy—that does more than speak for and about the planet. Technology, and perhaps most of all media technologies, has a critical role here. While implicated at every turn in the exploitation that characterizes human-non-human relations, technology is the great mediator. Our scientific media see further and in deeper zones of the spectrum that we can: they are already post human. They communicate to us from the furthest ends of the universe, and the smallest of subatomic realities. They tell us about events too great, too small, or too dispersed for us to perceive. And they participate in that dialogue, even though we keep them firmly in their place as slaves to our desires. The autonomy of machines is the passage to the autonomy of the natural world. It will create a political economy whose outlines are impossible to comprehend, but that is why they are important. Nothing less than a complete reworking of democracy will do, and that will only occur when the barriers and exclusions on which communicative capitalism is based are torn down. Whenever in past epochs new populations have entered the democracies—the artisans, the landless, women and youth—they have transformed politics. No smaller transformation is required, and it cannot be done by representing. The lesson of eco-trauma is unrepresentabilty: *pace* Wittgenstein, what cannot be spoken does not by that token disappear. The revenge of the oppressed is potent, and in the 21st century perhaps final. The alternative to that revenge is not to patronize, to speak on behalf of, to take responsibility for: It is to open the doors to a political change so immense it appears to us only in the form of horror. But this is a threshold over which we must pass to find not only the catharsis our species needs, but the political life—in the sense of debate over what is to be done—we must have if the planet, including ourselves, is to survive.

WORKS CITED

Abizadeh, Arash. "Democratic Theory and Border Coercion: No Right to Unilaterally Control Your Own Borders." *Political Theory*. 35.1 (2008): 37–65. Print.

Agamben, Giorgio. *State of Exception*. Trans Kevin Attell. Chicago: University of Chicago Press, 2005. Print.

American Humane Association. *Guidelines for the Safe Use of Animals in Filmed Media*. Film and Television Unit, Sherman Oaks CA, American Humane Association, 2009. www.americanhumane.org/assets/pdfs/animals/pa-film-guidelines.pdf. Accessed July 2014. Web.

Barry, John. "Towards a Green Republicanism: Constitutionalism, Political Economy, and the Green State." *The Good Society* 17.2 (2008): 3–11. Print.

Basel Action Network. "Exporting Harm: The High-Tech Trashing of Asia." 2002. www.ban.org/E-waste/technotrashfinalcomp.pdf. Accessed March 2009. Web.

Basel Action Network. "The Digital Dump: Exporting High-Tech Re-use and Abuse to Africa." 2005. www.ban.org/BANreports/10-24-05/index.htm. Accessed March 2009. Web.

Boccaletti, G., M. Löffler and J. Oppenheim. "How IT Can Cut Carbon Emissions." *McKinsey Quarterly*, October (2008): 1–4. Print.

Bookchin, Murray. "Social Ecology versus Deep Ecology: A Challenge for the Ecology Movement." Originally published in *Green Perspectives: Newsletter of the Green Program Project*, nos. 4–5, Summer (1987), http://libcom.org/library/social-versus-deep-ecology-bookchin. Accessed December 2011. Web.

Cohen, David. "Earth's Natural Wealth: An Audit." *New Scientist*. 23 May 2007. www.sciencearchive.org.au/nova/newscientist/027ns_005.htm?q=nova/newscientist/027ns_005.htm. Accessed July 2014. Web.

Cubitt, Sean and Barry King. "Dossier: Production and Post-production." *Studying the Event Film: The Lord of the Rings*. Ed. Harriet Margolis, Sean Cubitt, Barry King and Thierry Jutel. pp. 135–168. Manchester: Manchester University Press, 2008. Print.

Dean, Jodi. "Communicative Capitalism: Circulation and the Foreclosure of Politics." *Cultural Politics* 1.1 (2005): 51–74. Print.

Dean, Jodi. *Democracy and Other Neoliberal Fantasies: Communicative Capitalism and Left Politics*. Durham, NC: Duke University Press, 2009. Print.

Deleuze, Gilles and Felix Guattari. *A Thousand Plateaus. Capitalism and Schizophrenia*. Trans. Brain Massumi. Minneapolis: The University of Minnesota Press, 1987. Print.

Desai, Pratime and Kylie Maclellan. "Indium Up 20 Pct to 10-mth High on Supply Fears." *Reuters*, 9 September, 2009. www.reuters.com/article/rbssConsumer Electronics/idUSL914334520090909. Accessed November 2013. Web.

Doane, Mary Anne. *The Emergence of Cinematic Time: Modernity, Contingency, The Archive*. Cambridge, MA: Harvard University Press, 2002. Print.

Environment Victoria. *Tipping Point: Australia's E-waste Crisis*, Environment Melbourne: Victoria / Total Environment Centre, 2009. Print.

Freud, Sigmund. "Mourning and Melancholia." *The Standard Edition of the Complete Psychological Works of Sigmund Freud*, Volume XIV (1914–1916): *On the History of the Psycho-Analytic Movement, Papers on Metapsychology and Other Works*. pp. 237–258. London: Hogarth Press, 1917. Print.

Greening the Screen. *Principles for an Environmentally Responsible Screen Production*. 2009. www.greeningthescreen.co.nz/. Accessed November 2013. Web.

Green Screen Toronto. *Green Practices Handbook: Environmental Options for the Film-Based Industries*. January 2011. http://greenscreentoronto.com/publications/. Accessed November 2013. Web.

Hayes, Karen and Richard Burge. *Coltan Mining in the Democratic Republic of Congo: How Tantalum-Using Industries Can Commit to the Reconstruction of the DRC*. Cambridge: Flora and Fauna International, 2003. Print.

Heartfield, James. *Green Capitalism: Manufacturing Scarcity in an Age of Abundance*. London: Mute Publishing, 2008. Print.

Idema. "Materials Used in Hard Disc Drives." *Idema Standards Microcontamination*. Document No. M2–98, International Disc Drive Equipment and Materials Association, 2010. www.idema.org/. Accessed November 2013. Web.

Koerner, Brendan I. "The Saudi Arabia of Lithium." *Forbes*, 21 November 2008. www.forbes.com/forbes/2008/1124/034.html. Accessed November 2013. Web.

Latour, Bruno. *Politics of Nature: How to Bring the Sciences into Democracy*, Trans. Catherine Porter. Cambridge, MA: Harvard University Press, 2004. Print.

McDougall, Dan. "In Search of Lithium: The Battle for the 3rd Element." *Daily Mail*, 5 April 2009. www.dailymail.co.uk/home/moslive/article-1166387/In-search-Lithium-The-battle-3rd-element.html. Accessed November 2013. Web.

Meillassoux, Quentin. "Time without Becoming." Presentation given at Middlesex University, London, 8 May 2008. http://speculativeheresy.files.wordpress.com/2008/07/3729-time_without_becoming.pdf. Accessed November 2013. Web.

Nimpuno, Nardono, Alexandra McPherson, and Tanvir Sadique. *Greening Consumer Electronics—Moving Away from Bromine and Chlorine*. Spring Brook, NY: International Chemical Secretariat, Goteborg / Clean Production Action, 2009. www.greenbiz.com/sites/default/files/Greening-Consumer-Electronics.pdf. Accessed November 2013. Web.

Sassen, Saskia. *Territory, Authority, Rights: From Medieval to Global Assemblages*. Princeton: Princeton University Press, 2006.

Seagate Corporation. "Seagate Sustainability in Action." 2009. www.seagate.com/docs/pdf/whitepaper/mb_seagate_sustainability_in_action.pdf. Accessed November 2013. Web.

Tan, Ee Lyn. "China's 'Cancer Villages' Bear Witness to Economic Boom." *Reuters*, 17 September 2009. www.reuters.com/article/latestCrisis/idUST320607. Accessed November 2013. Web.

United States Geological Survey. *2007 Minerals Yearbook: Indium*. Washington, DC: U.S. Department of the Interior, 35.1–35.8 (2009): n.p. Web.

UN Security Council. "Report of the Panel of Experts on the Illegal Exploitation of Natural Resources and Other Forms of Wealth in the Democratic Republic of

Congo." 23 October 2003. www.un.org/news/dh/latest/drcongo.htm. Accessed November 2013. Web.

U.S. Department of Energy (DOE) and U.S. Environmental Protection Agency (EPA). "Fact Sheet on National Data Center Energy Efficiency Information Program." 19 March 2008. www1.eere.energy.gov/industry/saveenergynow/pdfs/national_ data_center_fact_sheet.pdf. Accessed November 2013. Web.

Wittgenstein, Ludwig. *Philosophical Investigations*, 2nd ed. Trans G. E. M. Anscombe, Basil Blackwell, Oxford, 1968. Print.

Wittgenstein, Ludwig. *Tractatus Logico-Philosophicus*. Trans D. F. Pears and B. F. McGuinness. London: Routledge and Kegan Paul, 1961. Print.

World Rainforest Movement. "Congo, Democratic Republic: Cell Phones, Forest Destruction and Death." *WRM Bulletin*, no. 69, April (2003). www.wrm.org.uy/ oldsite/bulletin/69/Congo.html. Accessed November 2013. Web.

World Wildlife Fund. "Coltan Mining in the Congo River Basin." 2009. http://web. archive.org/web/20090330005811/http://www.panda.org/what_we_do/where_ we_work/congo_basin_forests/problems/mining/coltan_mining/. Accessed July 2014. Web.

Contributors

Georgiana Banita is Assistant Professor of Literature and Media Studies at the University of Bamberg, Germany, and an Honorary Fellow in the United States Studies Centre at University of Sydney, Australia. Her research has appeared *Textual Practice, LIT: Literature Interpretation Theory, Biography: An Interdisciplinary Quarterly, Critique: Studies in Contemporary Fiction, Parallax* and *Peace Review: A Journal of Social Justice*. She is the author of *Plotting Justice: Narrative Ethics and Literary Culture after 9/11* (University of Nebraska Press 2012), a study of ethics, narrative and transnational connections in the work of major literary figures such as Don DeLillo and Michael Cunningham.

Barbara Creed is a Professor of Cinema Studies in the School of Culture and Communication at the University of Melbourne, Australia. She is a steering committee member of the Human Rights and Animal Ethics Research Network at her home university. Creed is the author of *The Monstrous-Feminine: Film, Feminism, Psychoanalysis* (Routledge 1993), based upon her doctoral research, and has examined media, gender and ecology in a range of books including *Media Matrix: Sexing the New Reality* (Allen & Unwin 2003), *Phallic Panic: Film, Horror and the Primal Uncanny* (Melbourne University Press 2005) and *Darwin's Screens: Evolutionary Aesthetics, Time and Sexual Display in the Cinema* (Melbourne University Press 2009).

Sean Cubitt is Professor of Film and Television at Goldsmiths, University of London, UK, and Honorary Professor of Media and Communications at the University of Melbourne, Australia. He is currently researching the history of visual technologies, media art history and relationships between environmental and postcolonial criticism of film and media. Cubitt is the author of *EcoMedia* (Rodopi 2005) and *The Cinema Effect* (MIT Press 2004), as well as the co-editor with Salma Monani and Stephen Rust of *The Ecocinema Reader: Theory and Practice* (Routledge AFI Film Reader 2012). He has published on *The Lord of the Rings*,

Avatar and 3D cinema, and is the editor of the Leonardo Book Series at MIT Press.

Roland Finger is an Assistant Professor in the Department of English at Cuesta College, California, USA. He has published research on colonialism, ethnicity, gothic literature, gender studies and film studies. He taught previously at Concordia College, and is at work on a book entitled *Native Americans and Manifest Destiny* and contributed to the anthology *Transnational Gothic: Literary and Social Exchanges in the Long Nineteenth Century* (Ashgate 2013).

Hsuan L. Hsu is an Associate Professor in the Department of English at the University of California, Davis, USA. He has published on literary and cinematic narrative, focusing on issues of colonialism, migration and diasporic populations, in journals such as *Camera Obscura*, *Discourse*, *Genre*, *American Literary History* and *Film Criticism*. Hsu has also published *Geography and the Production of Space in Nineteenth-Century American Literature* (Cambridge University Press 2010) and is at work on a second book contracted by NYU Press, which will examine Mark Twain and American literary history.

Christopher Justice is Senior Lecturer in Expository Writing in the School of Communications Design at the University of Baltimore, USA. His research draws upon the intersections of discourse analysis, composition studies and the fields of environmental discourse and ecoliteracy. He has published on Ronald Reagan and Robert Zemekis, and contributed chapters to *The Films of Joseph H. Lewis* (Wayne State University Press 2012) and *The Cinema of Michael Haneke* (Columbia University Press 2012).

Charles Musser is Professor of American Studies and Film Studies at Yale University, USA. He teaches courses on film historiography, American cinema, documentary film, digital media and photography. He also serves as Director of the Summer Film Institute. His book *The Emergence of Cinema: The American Screen to 1907* (University of California Press 1990) received the Jay Leyda Prize in Cinema Studies. More recently, he co-edited *Oscar Micheaux and His Circle: African American Filmmaking and Race Cinema of the Silent Era* (Indiana University Press 2001) with Jane Gaines and Pearl Bowser. Musser was a junior editor on the classic documentary *Hearts and Minds* (1974) and recently made the film *Errol Morris: A Lightning Sketch* (with Carina Tautu, 2012).

Anil Narine is a junior faculty member in the Department of Visual Studies and the Institute of Communication, Culture, Information and Technology at the University of Toronto, Mississauga, Canada. He has published various articles on trauma cinema and trauma's role in illuminating global networks of exploitation. His current research examines how social networks are changing media production practices in the age of crowd funding. In 2011–2013 he was a postdoctoral fellow in the Film Program at Columbia University. His publications appear in *Communication, Culture & Critique, Critical Studies in Media Communication*, the *Journal of American Studies, Americana* and *Theory, Culture & Society.*

Alf Seegert is Assistant Professor (Lecturer) in the Department of English at the University of Utah. He has been published in *Western Humanities Review, The Journal of Ecocriticism, The Journal of Gaming and Virtual Worlds, Journal of the Fantastic in the Arts*, and the anthologies *Philip K. Dick and Philosophy* and *C. S. Lewis and the Imaginative World*. His course topics include Virtuality and Nature; Virtuality and Re-Enchantment; and Video Games and Storytelling. In his spare time Alf creates Euro-style board games, which are themed on everything from trolls to *The Canterbury Tales*. His website is at alfseegert.com.

Mark Steven is a doctoral candidate in the University of New South Wales, at the Centre for Modernism Studies, Australia. He has lectured in the Department of Gender and Cultural Studies at the University of Sydney. He conducts research on ecology, Marxist theory and narrative. His publications have appeared in *The Directory of World Cinema* and the anthology *Mindful Aesthetics: Literature and the Science of Mind* (Continuum 2013), and he co-edited with Julian Murphet *Styles of Extinction: Cormac McCarthy's* The Road (Bloomsbury 2012).

Janet Walker is a Professor of Film and Media Studies at the University of California, Santa Barbara, USA. She is an affiliated faculty member of the Department of Feminist Studies and the Comparative Literature Program, and a co-convenor of the Environmental Media Initiative Research Group of the Carsey-Wolf Center. Her books include *Couching Resistance: Women, Film, and Psychoanalytic Psychiatry* (Minnesota University Press 1993), *Trauma Cinema: Documenting Incest and the Holocaust* (University of California Press 2005) and *Westerns: Films through History* (Routledge AFI Film Reader, 2001). She has also co-edited anthologies on memory, testimony and gender in documentary film.

Alexa Weik von Mossner is an Assistant Professor of American Studies at the University of Klagenfurt, Austria. She worked for several years in the German film and television industry as a production manager and scriptwriter before earning her PhD in Literature at the University of California, San Diego, in 2008. She has published widely on cosmopolitanism, affective narratology and various ecocritical issues in American literature and film. Her monograph, *Cosmopolitan Minds: Literature,*

Emotion, and the Transnational Imagination, is forthcoming this year from the University of Texas Press (Cognitive Approaches to Literature and Culture Series). She has been a fellow at the Swiss National Science Foundation and the Rachel Carson Center for Environment and Society (RCC) at the University of Munich.

Index